Y0-BSN-510

About Island Press

Since 1984, the nonprofit Island Press has been stimulating, shaping, and communicating the ideas that are essential for solving environmental problems worldwide. With more than 800 titles in print and some 40 new releases each year, we are the nation's leading publisher on environmental issues. We identify innovative thinkers and emerging trends in the environmental field. We work with world-renowned experts and authors to develop cross-disciplinary solutions to environmental challenges.

Island Press designs and implements coordinated book publication campaigns in order to communicate our critical messages in print, in person, and online using the latest technologies, programs, and the media. Our goal: to reach targeted audiences—scientists, policymakers, environmental advocates, the media, and concerned citizens—who can and will take action to protect the plants and animals that enrich our world, the ecosystems we need to survive, the water we drink, and the air we breathe.

Island Press gratefully acknowledges the support of its work by the Agua Fund, Inc., Annenberg Foundation, The Christensen Fund, The Nathan Cummings Foundation, The Geraldine R. Dodge Foundation, Doris Duke Charitable Foundation, The Educational Foundation of America, Betsy and Jesse Fink Foundation, The William and Flora Hewlett Foundation, The Kendeda Fund, The Andrew W. Mellon Foundation, The Curtis and Edith Munson Foundation, Oak Foundation, The Overbrook Foundation, the David and Lucile Packard Foundation, The Summit Fund of Washington, Trust for Architectural Easements, Wallace Global Fund, The Winslow Foundation, and other generous donors.

The opinions expressed in this book are those of the author(s) and do not necessarily reflect the views of our donors.

A Manual for Assessment Practitioners

Ecosystems and Human Well-being

A Manual for Assessment Practitioners

Neville Ash
Hernán Blanco
Claire Brown
Keisha Garcia
Thomas Henrichs
Nicolas Lucas
Ciara Raudsepp-Hearne
R. David Simpson
Robert Scholes
Thomas P. Tomich
Bhaskar Vira
Monika Zurek

Washington | Covelo | London

Library of Congress Cataloging-in-Publication Data

Ecosystems and human well-being : a manual for assessment practitioners / Neville Ash ... [et al.].
p. cm.
Includes bibliographical references and index.
ISBN 978-1-59726-711-3 (pbk. : alk. paper) —
ISBN 978-1-59726-710-6 (cloth : alk. paper)
1. Human ecology. 2. Ecosystem management. 3. Biodiversity. I. Ash, Neville.
GF50.E2617 2010
333.71'4—dc22 2009043012

Printed on recycled, acid-free paper

Manufactured in the United States of America

10 9 8 7 6 5 4 3 2 1

Contents

Foreword

Commissioned by the United Nations Secretary–General in 2000, and completed in 2005, the Millennium Ecosystem Assessment (MA), based on the findings of 34 "sub-global" assessments carried out in a diverse set of ecosystems in sites around the world, provides a state-of-the-art appraisal of the condition and trends in the world's ecosystems and the services they provide.

The MA presents compelling evidence that underlines the urgency and necessity of restoring, conserving, and sustainably managing our ecosystems. Most important, the assessment shows that, with appropriate actions, it is possible to reverse the degradation of many ecosystem services over the next 50 years. By providing invaluable information to policy makers, the MA seeks to help ensure that the required changes in current policy and practice undertaken will be evidence based and informed by the best available scientific analysis.

This manual, *Ecosystems and Human Well-being: A Manual for Assessment Practitioners*, allows for the wider adoption of the MA conceptual framework and methods. The manual, which contains numerous case studies of best practice, offers a practical guide for undertaking ecosystem assessments and includes tools and approaches that can assess options for better managing ecosystems.

UNEP and UNDP, working together with other partners, are committed to promoting sustainable development and ensuring the protection of our planet. By stimulating future ecosystem assessments, based on the proven methodologies of the MA, it is our hope that this manual will provide the knowledge needed to develop appropriate and effective policies and strategies to ensure that the earth's ecosystems and their vital services are restored and preserved. Our very livelihoods depend on this.

Achim Steiner, Executive Director
United Nations Environment
Programme (UNEP)

Helen Clark, Administrator
United Nation Development Programme
(UNDP)

Preface

The Millennium Ecosystem Assessment

The Millennium Ecosystem Assessment (MA) was called for by the United Nations Secretary-General Kofi Annan in 2000 in his report to the UN General Assembly, *We the Peoples: The Role of the United Nations in the 21st Century*. The MA was carried out between 2001 and 2005 to assess the consequence of ecosystem change for human well-being, by attempting to bring the best available information and knowledge on ecosystem services to bear on policy and management decisions. The MA established the scientific basis for action needed to enhance the conservation and sustainable use of ecosystems and their contribution to human well-being. The MA was in part a global assessment, but to facilitate better decision making at all scales, 34 regional, national and local scale assessments (or sub-global assessments) were included as core project components. Since the release of the MA, further sub-global assessments have started.

What are ecosystems and ecosystem services?

An *ecosystem* is a dynamic complex of plant, animal and micro-organism communities and the nonliving environment interacting as a functional unit. The conceptual framework for the MA assumes that people are integral parts of ecosystems. The MA Report itself focuses on linkages between ecosystems and human well-being, in particular on "ecosystem services," which are the benefits that people obtain from ecosystems.

Ecosystem services include the following:

- ***Provisioning services,*** such as providing food, water, timber and fibre;
- ***Regulating services,*** such as the regulation of climate, floods, disease, wastes and water quality;
- ***Cultural services,*** such as offering recreational, aesthetic, and spiritual benefits; and
- ***Supporting services,*** such as soil formation, photosynthesis, and nutrient cycling.

What is ecosystem assessment?

An ecosystem assessment provides the connection between environmental issues and people. An assessment of ecosystem services needs to consider both the ecosystems from which the services are derived and also the people who depend on and are affected by changes in the supply of services, thereby connecting environmental and development sectors. Assessments play numerous roles in the decision-making process, including responding to decision makers' needs for information, highlighting trade-offs between decision options, and modeling future prospects to avoid unforeseen long-term consequences. They inform decisions by providing critical judgment of options and uncertainty and through synthesizing and communicating complex information on relevant issues. They are also of value through the process they involve, engaging and informing decision makers long before final assessment products are available.

Thus, decision makers—including those whose goals and actions are focused on people, society, and economics—can benefit from examining the extent to which achieving their goals depends on ecosystem services. Assessments can provide credible and robust information on the links between ecosystems and the attainment of economic and social goals.

Why is this Manual needed?

This Manual makes the methods of the MA and associated sub-global (local and regional) assessments widely accessible. While the MA is the most comprehensive assessment of ecosystems carried out to date, there are other related assessment processes such as Global Environment Outlook (GEO), Global International Waters Assessment (GIWA), Intergovernmental Panel on Climate Change (IPCC), Land Degradation Assessment in Drylands (LADA), International Assessment of Agricultural Knowledge, Science and Technology for Development (IAASTD) and World Water Assessment. Lessons learned from these assessments supplement the best practice of ecosystem assessment identified through the MA. The publication of this Manual aims to encourage more assessments at scales which are relevant to policy and decision makers.

Why use this Manual?

The Manual is intended to be a "how to" guide for undertaking ecosystem assessments. The Manual contains detailed guidance on conceptual frameworks, assessing status and trends of ecosystems, developing and using scenarios, assessing policy options, and the process for establishing, designing and running an ecosystem assessment, including communications and outreach.

The priority audience for the Manual are individuals who are responsible for designing and carrying out environmental or developmental assessments, and individuals responsible for building capacity for ecosystem assessments, either through structured training (such as through developing curricula relating to ecosystem services and development) or assistance in conducting assessments on the ground.

New and emerging ecosystem assessment practitioners should use this Manual to:

- Familiarize themselves with the concept of ecosystem assessment;
- Understand how and why an ecosystem assessment can benefit decision making at their scale of interest and what steps are involved;
- Improve capacity to undertake an assessment where the need for one has already been identified; and
- Act as a guide for practitioners who are undertaking an assessment to obtain more background information and identify sources of potential assistance with challenging areas.

Experienced ecosystem assessment practitioners should use this Manual to:

- Update and complement their knowledge and skills in ecosystem assessment;
- Serve as a basis for dialogue on methods for ecosystem assessment to improve the shared knowledge base on this approach; and
- Train new and emerging ecosystem assessment practitioners in an applied or classroom setting.

This Manual complements other related resources such as the *Ecosystem Services: A Guide for Decision-makers,* prepared by WRI and others (which focuses on how the findings of ecosystem assessments can be used), and the UNEP-GEO assessment training modules (which cover a much broader institutional State of the Environment reporting process).

Acknowledgments

Authors of the Millennium Ecosystem Assessment (MA) were the recipients of the Zayed International Prize for the Environment prize in honor of His Highness Sheik Zayed Bin Sultan Al Nahyan, President of the United Arab Emirates and Governor of Abu Dhabi. On agreement of the Board of the MA, a significant portion of these funds was allocated to the production of this Manual. Financial support for the Manual was also provided by the UK Department for Environment, Food and Rural Affairs.

The support of the following institutions enabled the participation of lead authors in compiling this Manual: IUCN, Switzerland; RIDES, Chile; The Cropper Foundation, Trinidad and Tobago; National Environmental Research Institute, Denmark; European Environment Agency (EEA), Denmark; Secretariat of Environment and Sustainable Development, Argentina; McGill University, Canada; Council for Scientific and Industrial Research, South Africa; National Center for Environmental Economics, USA; University of California, Davis, USA; University of Cambridge, UK; FAO, Italy; and United Nations Environment Programme World Conservation Monitoring Centre (UNEP-WCMC), UK.

In addition to the lead authors, contributions to the Manual were provided by Alejandro Argumedo (Asociacion Andes), Ivar Baste (UNEP), Karen Bennett (WRI), Esther Camac (Association IXACAVAA for Indigenous Development and Information), Anantha Duraiappah (UNEP), Bas Eickhout (Netherlands Environment Assessment Agency), Colin Filer (Australian National University), Kelly Garbach (University of California, Davis), Helmut Geist (University of Aberdeen), Habiba Gitay (World Bank Institute), Kasper Kok (University of Wageningen), Frances Irwin (WRI), Anne-Marie Izac (Consultative Group for International Agricultural Research), Louis Lebel (Chiang Mai University), Marcus Lee (The World Bank), Walter Reid (David and Lucile Packard Foundation), Maiko Nishi (United Nations University), Meine van Noordwijk (World Agroforestry Centre), Lennart Olsson (Lund University), Cheryl Palm (Columbia University), Janet Ranganathan (WRI), Maurice Rawlins (The Cropper Foundation), Teresa Ribeiro (European Environment Agency), Axel Volkery (European Environment Agency), and Detlef van Vuuren (Netherlands Environment Assessment Agency).

All of the MA authors and review editors who contributed to this manual through their contributions to the MA itself, upon which this manual is based, are acknowledged. The MA process included a total of 34 sub-global assessments (SGAs) from around the world. These assessments analyzed the importance of ecosystem services for human well-being at local, national, and regional scales. The experience of the team involved in carrying out or coordinating SGAs also contributed to the content reflected in the manual.

The Manual Chapter Review Editors provided valuable guidance and input: John Agard (University of West Indies); Yolanda Kakabadze (Fundación Futuro Latinoamericano), Richard Norgaard (University of California, Berkeley), David Stanners (European Environment Agency), Dan Tunstall (WRI) and Tom Wilbanks (Oak Ridge National Laboratory).

Many individuals reviewed drafts of the Manual and provided critical review comments: Heidi Albers (Oregon State University), Salvatore Arico (UNESCO), Jan Bakkes (Netherlands Environmental Assessment Agency), Karen Bennett (WRI), Reinette Biggs (Stockholm Resilience Institute), Traci Birge (UNEP DEWA), Steve Carpenter (University of Wisconsin), David Cooper (Secretariat for the Convention on Biological Diversity), Steven Cork (EcoInsights), Anantha Duraiappah (UNEP), Paul Ferraro (Georgia State University), Max Finlayson (Charles Sturt University), Inger Heldal (Norad—Norwegian Agency for Development Cooperation), Bente Herstad (Norad—Norwegian Agency for Development Cooperation), Robert Höft (Secretariat for the Convention on Biological Diversity), Frances Irwin (WRI), Annekathrin Jaeger (EEA), Christian Layke (WRI), Marcus Lee (UNEP, now World Bank), Markus Lehmann (Secretariat for the Convention on Biological Diversity), Jock Martin (EEA), Connie Musvoto (CSIR), Gerald Nelson (University of Illinois at Urbana-Champaign), David Niemeijer, Cheryl Palm (The Earth Institute, Columbia University), Bely Pires, Janet Ranganathan (WRI), Walter Reid (David and Lucile Packard Foundation), Belinda Reyers (CSIR), Dale Rothman (International Institute for Sustainable Development), Frederik Schutyser (EEA), Albert van Jaarsveld (National Research Foundation, South Africa), Hans Vos (EEA), Rodrigo Victor, Ernesto Viglizzo, Angela Wilkinson (James Martin Institute, Oxford University), Sven Wunder (CIFOR), Kakuko Nagatani Yoshida (UNEP Regional Office for Latin America and the Caribbean) and the participants in the first meeting of sub-global assessment practitioners held in Kuala Lumpur, April 2008.

Translation of the manual into French was sponsored by the UNDP/UNEP Poverty and Environment Initiative. The Spanish translation was sponsored by the European Environment Agency.

UNEP-WCMC coordinated the preparation and publication of the manual. Thanks go to Claire Brown, Matt Walpole, Philip Bubb, and Jessica Jones for their efforts. Keisha Garcia from the Cropper Foundation was hosted as a fellow at UNEP-WCMC and advanced the manual during her tenure and beyond. Linda Starke's expert eye to the text as copyeditor has been an invaluable contribution.

Acronyms and Abbreviations

AC	Advisory Committee
AKST	agricultural knowledge, science and technology
AR4	Fourth Assessment Report (for the IPPC)
ASB	the Alternatives to Slash-and-Burn Program
ATEAM	Advanced Terrestrial Ecosystem Analysis and Modelling
BBOP	the Business Biodiversity Offset Program
CA	California Agroecosystem Assessment
CBA	cost-benefit analysis
CBD	Convention on Biological Diversity
CEA	cost-effectiveness analysis
CEC	Cation exchange capacity
CEO	chief executive officer
CFCs	Chlorofluorocarbons
COP	Conference of the Parties
CRP	Conservation Reserve Program
DAC	Development Assistance Committee
DDP	Dahlem Desertification Paradigm
DfID	U.K. Department for International Development
DPSIR	Drivers-Pressures-States-Impacts-Responses
EC	the European Commission
EEA	European Environment Agency
GDP	Gross Domestic Product
GEO	Global Environment Outlook
GIS	Geographic Information System
GIWA	Global International Waters Assessment
GNI	Gross National Income
HIPC	Heavily Indebted Poor Countries
HWB	Human Well-Being
IAASTD	International Assessment of Agricultural Science and Technology for Development
IEA	International Energy Agency
ILO	International Labour Organization
IMF	International Monetary Fund
INRM	Integrated natural resource management
IPCC	intergovernmental panel on climate change
IPU	Inter-Parliamentary Union

ITU	International Telecommunication Union
IUCN	International Union for Conservation of Nature
JIU	Joint Inspection Unit
LADA	Land Degradation Assessment in Drylands
LDCs	Least Developed Countries
LLDA	Laguna Lake Development Authority
LUCC	Land-Use/Cover Change Programme
MA	Millennium Ecosystem Assessment
MBIs	market-based incentives
MCA	multi-criteria analysis
MDGs	Millennium Development Goals
NGO	nongovernmental organization
ODA	Official Development assistance
ODP	Ozone depletion potential
OECD	Organisation for Economic Co-operation and Development
PES	payments for ecosystem services
PNG	Papua New Guinea
PPP	purchasing power parity
PRELUDE	PRospective Environmental analysis of Land Use Development in Europe
PSR	Pressure State Response
SAfMA	Southern Africa Millennium Ecosystem Assessment
SDM	summary for decision makers
SEWA	Self-Employed Women's Association
SPM	summary for policy makers
SRES	Special Report on Emissions Scenarios
STEEP	social-cultural, technological, economic, environmental, and political
drivers	driving forces
ToR	terms of reference
UN	United Nations
UNAIDS	The Joint United Nations Programme on HIV/AIDS
UNCCD	United Nations Convention to Combat Desertification
UNCTAD	United Nations Conference on Trade and Development
UNDP	United Nations Development Programme
UNEP	United Nations Environment Programme
UNESCO	United Nations Educational, Scientific and Cultural Organization
UNFCCC	United Nations Framework Convention on Climate Change
UNICEF	United Nations Children's Fund
UNSD	United Nations Statistic Division
UPLB	University of the Philippines Los Baños
U.S.	United States of America
WHO	World Health Organisation
WRI	World Resources Institute
WTO	World Trade Organisation

1

Assessing Ecosystems, Ecosystem Services, and Human Well-being

Neville Ash, Karen Bennett, Walter Reid, Frances Irwin, Janet Ranganathan, Robert Scholes, Thomas P. Tomich, Claire Brown, Habiba Gitay, Ciara Raudsepp-Hearne, and Marcus Lee

---

What is this chapter about?

This chapter provides an overview of the process and components of scientific assessments that have as their focus or include within their scope the connections between ecosystems and people. It introduces ecosystem services as the link between ecosystems and human well-being and therefore as the focus of assessing the consequences of ecosystem changes for people. The chapter introduces and highlights the relationship between the various components of assessment. In doing so, it provides an introduction and roadmap to the subsequent chapters of the manual.

1.1 Introduction

Section's take-home messages

- This manual can be used as a whole document, or individual chapters can help assessment practitioners who are looking for guidance on particular aspects of the process.
- Assessments are not just about the findings. Getting the process right, from the early stages of design through to the communication of findings, is essential in order to have an impact.

This manual is a stand-alone "how-to" guide about conducting an assessment of the consequences of ecosystem change for people. However, the manual also relates closely to other recent publications, particularly *Ecosystem Services: A Guide for Decision Makers* (WRI 2008), which presents methods for public-sector decision makers to use information on ecosystem services to strengthen economic and social development policies and strategies. This manual can be used as a whole document, or individual chapters can help assessment practitioners who are looking for guidance on particular aspects of the process. The manual builds on the experiences and lessons learned from the Millennium Ecosystem Assessment (MA) global assessment and from over 30 ongoing or completed sub-global assessment initiatives at a range of scales, including local, national, and regional assessments. (See www.MAweb.org for further details on the MA and the various follow-up activities currently under way.) It also includes insight and experiences gained from a wider range of assessment activities focused on ecosystem services.

The chapter begins with an overview of such assessments—what they are and why they are useful—and then provides a summary of the step-by-step process for conducting an assessment. Drawing on both theory and best practice from the field and on a range of global and sub-global assessments, the chapter highlights the importance not just of the findings of an assessment but also of the process itself. Getting the process right, from the early stages of design through to the communication of findings, is essential in order to have an impact on the intended audience.

This manual has been written to support integrated ecosystem assessment practitioners. However, it is essential that the assessment practitioner also understand the decision-making context in which the study is being conducted and into which the findings may be taken on board. As such, the chapter concludes with a short section on how assessments can be considered in the context of the decision-making process and how the focus and impact of an assessment will depend on what stage an issue is in its policy life cycle.

Subsequent chapters in the manual elaborate on the material presented here and address key aspects of the assessment process: engaging stakeholders; developing and using a conceptual framework; conducting assessments of conditions and trends in ecosystems, their services, and human well-being; developing scenarios of change for ecosystems, their services, and human well-being; and assessing responses or interventions that aim to improve the management of ecosystems for people. Figure 1.1 outlines the main contents and layout of this manual, and shows how key sections of the manual relate.

1.2 How to improve decision making using ecosystem assessments

---

Section's take-home messages

- An ecosystem services assessment can help build a bridge between the development and environmental communities by providing credible and robust information on the links between ecosystem management and the attainment of economic and social goals.
- As improvements are made in describing and valuing the benefits of ecosystem services, decision makers can better understand how their actions might change these services, consider the trade-offs among options, and choose policies that sustain the appropriate mix of services.
- Successful assessments share three basic features: they are credible, legitimate, and relevant to decision makers' needs.

People everywhere depend on ecosystems for their well-being. Ecosystems are the source of obvious necessities such as food and fresh water, but they also provide

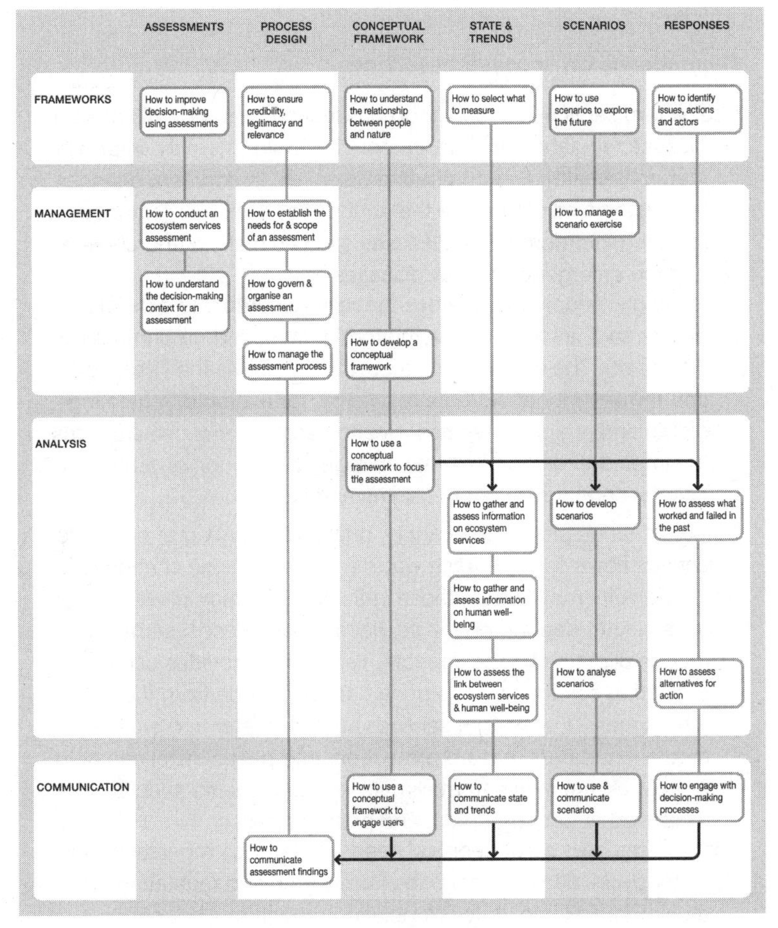

Figure 1.1. Contents and layout of the manual.

less obvious services such as flood protection, pollination, and the decomposition of organic waste. The natural world provides spiritual and recreational benefits as well. These and other benefits of the world's ecosystems have supported the extraordinary growth and progress of human societies. Yet the Millennium Ecosystem Assessment found that the majority of ecosystem services are in a state of decline and can no longer be taken for granted. Ignoring the links between ecosystems and human well-being in public and private decision making puts at risk our ability to achieve long-term development goals. An assessment of ecosystem services provides the connection between environmental issues and people. Thus, decision makers—including those whose goals and actions are focused on people, society, and economics—can benefit from examining the extent to which achieving their goals depends on ecosystem services (see Table 1.1).

Reconciling economic development and nature is challenging because they have traditionally been viewed in isolation or even in opposition, and the full extent of humanity's dependence on nature's benefits, or ecosystem services, is seldom taken into account by development or environmental communities. An ecosystem services

Table 1.1. Linking development goals and ecosystem services

Goal	Dependence on ecosystem services
Health	Ecosystem services such as food production, water purification, and disease regulation are vital in reducing child mortality, improving maternal health, and combating diseases. In addition, changes in ecosystems can influence the abundance of human pathogens, resulting in outbreaks of diseases such as malaria and cholera and the emergence of new diseases.
Natural hazard protection	Increasingly, people live in areas that are vulnerable to extreme events such as floods, severe storms, fires, and droughts (MA 2005:443). The condition of ecosystems affects the likelihood and the severity of extreme events by, for example, regulating global and regional climates. Healthy ecosystems can also lessen the impact of extreme events by regulating floods or protecting coastal communities from storms and hurricanes.
Adaptation to climate change	Climate change alters the quantity, quality, and timing of ecosystem service flows such as fresh water and food. These changes create vulnerabilities for those individuals, communities, and sectors that depend on the services. Healthy ecosystems can reduce climate change impacts. Vegetation provides climate-regulating services by capturing carbon dioxide from the atmosphere. Ecosystem services such as water and erosion regulation, natural hazard protection, and pest control can help protect communities from climate-induced events such as increased floods, droughts, and pest outbreaks.
Freshwater provision	Ecosystems help meet peoples' need for water by regulating the water cycle, filtering impurities from water, and regulating the erosion of soil into water. Population growth and economic development have led to rapid water resource development, however, and many naturally occurring and functioning systems have been replaced with highly modified and human-engineered systems. Needs for irrigation, domestic water, power, and transport are met at the expense of rivers, lakes, and wetlands that offer recreation, scenic values, and the maintenance of fisheries, biodiversity, and long-term water cycling.
Environmental conservation	Conservation projects often only consider a few benefits of nature's preservation. An ecosystem services framework can help build support for these projects by clarifying that their success provides multiple ecosystem services and therefore is linked to the achievement of other development goals. If a protected area, for example, can be shown to have additional benefits such as providing biochemicals for pharmaceuticals, its creation is more likely to be supported.

Table 1.1. continued

Goal	Dependence on ecosystem services
Food production	Ecosystems are vital to food production, yet there is pressure to increase agricultural outputs in the short term at the expense of ecosystems' long-term capacity for food production. Intensive use of ecosystems to satisfy needs for food can erode ecosystems through soil degradation, water depletion, contamination, collapse of fisheries, or biodiversity loss.
Poverty reduction	The majority of the world's 1 billion poorest people live in rural areas. They depend directly on nature for their livelihoods and well-being: food production, freshwater availability, and hazard protection from storms, among other services. Degradation of these services can mean starvation and death. Investments in ecosystem service maintenance and restoration can enhance rural livelihoods and be a stepping stone out of poverty.
Energy security	Many renewable energy sources, such as biofuels or hydroelectric power, are derived from ecosystems and depend on nature's ability to maintain them. Hydropower, for example, relies on regular water flow as well as erosion control, both of which depend on intact ecosystems.

Source: WRI 2008.

assessment can help build a bridge between the development and environmental communities by providing credible and robust information on the links between ecosystem management and the attainment of economic and social goals. This can mean the difference between a successful strategy and one that fails because of an unexamined consequence, for example for a freshwater supply, an agricultural product, a sacred site, or another ecosystem service (see Box 1.1).

Undertaking an ecosystem services assessment and taking the findings into account in policies and action can improve the long-term outcome of decisions. As improvements are made in describing and valuing the benefits of ecosystem services, decision makers can better understand how their actions might change these services, consider the trade-offs among options, and choose policies that sustain the appropriate mix of services. A range of assessment initiatives in recent years have focused on various aspects of ecosystem services. Box 1.2 provides an overview of the main recent and ongoing global assessment initiatives; further resources and background information on ecosystem services can be found in the "Additional Resources" section at the end of this chapter.

An assessment of ecosystem services needs to consider both the ecosystems from which the services are derived and also the people who depend on and are affected by changes in the supply of services, thereby connecting environmental and development sectors. Assessments play numerous roles in the decision-making process, including responding to decision makers' needs for information, highlighting trade-offs between decision options, and analyzing ecosystems to avoid unforeseen long-term consequences. They inform decisions through providing critical judgment

Box 1.1. The trade-off between food and fuel

Global food prices have been on the rise since 2000; they rose nearly 50 percent in 2007 alone. The price of basic staples, such as corn, oilseed, wheat, and cassava, is predicted to increase 26–135 percent by 2020. The recent increase in the cost of grain-based staples, such as tortillas in Mexico, beef noodles in western China, and bread in the United States, has several causes, including the emerging consequence of the increase in bioenergy production.

Promoted as a clean, sustainable alternative to fossil fuels, industrial countries have set increasingly higher mandates for the use of bioenergy to combat global climate change. Efforts to meet the rapid increase in demand for bioenergy have led to a global competition for limited natural resources such as land and water. Experts predicted that 30 million extra tons of corn—half of the global grain stock—would be dedicated to ethanol production in 2008 in the United States. On average, the grain required to make enough ethanol to fill a large car is enough to feed a person for a whole year. Crops can be used as food or fuel; both are important ecosystem services provided by nature. As countries continue to target corn and other agricultural products as the future supply of fuel, however, less food becomes available and food prices increase worldwide.

While the potential benefits of bioenergy can range from lower greenhouse gas emissions (in some cases) to renewability and energy independence, there are often trade-offs across other ecosystem services as a result of increased biofuel production. In addition to decreased food supply, the possible trade-offs include water quality impacts associated with increases in aggregate fertilizer use, nutrient runoff and erosion, and in some cases an increase in greenhouse gas emissions. Although the economic, social, and environmental effects of the recent biofuel push are not yet fully understood, many countries are rapidly expanding the area dedicated to these crops. It is in situations such as this that policy makers can benefit enormously from thorough assessments to determine options and better understand the consequences of their decisions.

of options and uncertainty and through synthesizing and communicating complex information on relevant issues. They are also of value through the process they involve, which engages and informs decision makers long before final assessment products are available.

Successful assessments share three basic features:

- *First, they are credible.* Involving eminent and numerous scientists as authors and expert reviewers and ensuring that all reports undergo expert peer review will help to ensure credibility. Assessments should focus not only on what is known with certainty by the scientific community but also on what remains uncertain. The clarity that assessments have given to areas of real scientific uncertainty (such as climate change in the 1990s) has been just as important in guiding policy as the clarity they have provided where there is broad scientific agreement. Moreover, by identifying areas of scientific uncertainty that matter for policy decisions (e.g., the ability to predict thresholds of change in socioecological systems), assessments can also help stimulate more support for scientific research.

Box 1.2. Recent and ongoing international assessments that focus on ecosystem services

Millennium Ecosystem Assessment

The Millennium Ecosystem Assessment, released in 2005, assessed the consequences of ecosystem change for human well-being. The MA consisted of a global assessment and 34 subglobal assessments to assess current knowledge on the consequences of ecosystem change for people. The MA brought about a new approach to assessment of ecosystems: a consensus of a large body of social and natural scientists, the focus on ecosystem services and their link to human well-being and development, and identification of emergent findings. The MA findings highlight the strain that human actions are placing on the rapidly depleting ecosystem services but also that appropriate action through policy and practice is possible. (www.MAweb.org)

International Assessment of Agricultural Science and Technology for Development

The International Assessment of Agriculture Science and Technology for Development (IAASTD), released in 2008, was an intergovernmental process that evaluated the relevance, quality, and effectiveness of agricultural knowledge, science, and technology (AKST) and the effectiveness of public- and private-sector policies as well as institutional arrangements in relation to AKST. The IAASTD consisted of a global assessment and five subglobal assessments using the same assessment framework, focusing on how hunger and poverty can be reduced while improving rural livelihoods and facilitating equitable, environmental, social, and economical sustainable development through different generations and increasing access to and use of agricultural knowledge, science, and technology. (www.agassessment.org)

Intergovernmental Panel on Climate Change

The Intergovernmental Panel on Climate Change (IPCC) released its fourth report (AR4) in 2007. The IPCC was established to provide decision makers with an objective source about climate change. Similar to the MA, the IPCC does not conduct any research or monitor specific data and parameters; it assesses the latest scientific, technical, and socioeconomic literature in an objective, open, and transparent manner. Ecosystem services are addressed in the fourth report of the IPCC by the reports of Working Group II (Impacts, Adaptation and Vulnerability) and Working Group III (Mitigation of Climate Change). The findings of AR4 highlighted a number of overarching key issues in relation to ecosystems and the services they provide for climate mitigation and adaptation. Specifically, the report drew links between the loss of ecosystem services and the reduction of societal option for adaptation responses. (www.ipcc.ch)

Land Degradation Assessment of Drylands

The Land Degradation Assessment of Drylands (LADA) is an ongoing assessment that aims to assess causes, status, and impact of land degradation in drylands in order to improve decision making for sustainable development at local, national, subregional, and global levels. Currently the LADA is focusing on developing tools and identifying available data that will be required to discover status and trends, hotspots of degradation, and bright spots (where degradation has been slowed or reversed). (http://lada.virtualcentre.org/)

Global Environment Outlook

The Global Environment Outlook (GEO) is the United Nations Environment Programme's ongoing assessment of the environment globally. The fourth GEO was released in 2007 and consists of a global assessment and subglobal assessments. GEO-4 provides information for decision makers on environment, development, and human well-being. (www.unep.org/geo)

- *Second, they are legitimate.* It is relatively easy for the administration of one province or country to ignore an assessment and report done by experts in another province or country, or for the CEO of a private company to ignore the findings of a report by a nongovernmental organization (NGO). What possible leverage would such a report have, no matter how thorough the science in it? Thousands of assessments and studies are published every year; what gives an assessment more weight with decision makers than others? Partly it is the authoritative status and credibility of the assessment through the organizations and individuals involved. But, equally important, the involvement of users of the assessment in the process itself ensures greater impact with decision makers through instilling a sense of "ownership" of the findings. A successful assessment is one that is legitimate in the eyes of the users, where decision makers use it as their own product.
- *Third, they are relevant (or salient) to decision makers' needs.* This is not to say that scientists do not have an opportunity to introduce new issues and findings that decision makers need to be aware of. They certainly do. But the priority for the assessment is to inform decisions that are being faced or soon will be faced by decision makers, at a particular scale, and in a particular context. (Section 1.4.2 provides further details on ensuring an assessment is policy relevant.)

Early on, assessments should evaluate whether they meet these three criteria and, if not, take the steps necessary to incorporate them. Chapter 2 explores in greater detail the various approaches to implementing these criteria and elaborates on the importance of stakeholder involvement at all stages in the assessment process.

1.3 How to conduct an ecosystem assessment—an overview

Section's take-home messages

- The assessment process has three key stages that are generally sequential but usually overlapping and iterative: the exploratory stage, the design stage, and the implementation of the assessment workplan.
- In the design stage, assessment organizers need to consider governance of the process, conceptual frameworks, how to link the different scales to be addressed, how to bridge different knowledge systems, capacity-building needs, and how to evaluate the process.
- The implementation stage requires the greatest need for flexibility. This stage generally assesses conditions and trends in ecosystems and their services, scenarios for the future, and past and current responses taken to enhance the contribution of ecosystems to human well-being.

An assessment will have the greatest impact where consideration is given to both process and products, where stakeholders are fully engaged, and where assessment design follows scoping of user needs. The assessment process has three key stages that are generally sequential but usually overlapping and iterative: the exploratory stage, the design stage, and the implementation of the assessment workplan (see

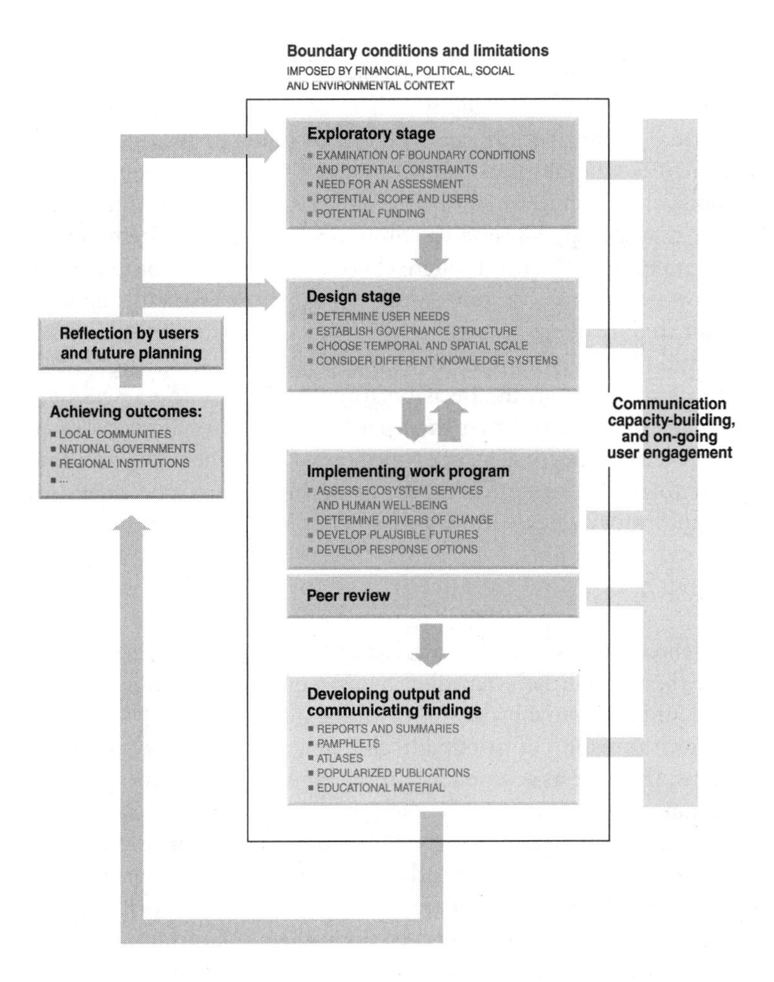

Figure 1.2. An overview of the assessment process.

Figure 1.2). User engagement, communication, and capacity building occur throughout the entire assessment process. A review process is essential for the assessment and provides both credibility and an opportunity for further engagement of users. All assessments are likely to need to be flexible and adaptive—to changing circumstances, user requirements, and process (and funding) constraints.

1.3.1 Exploratory stage

Determining the need for an assessment

The concept of an authorizing environment is a useful way to ensure that an assessment has the necessary level of buy in from key stakeholders. The authorizing environment is the set of institutions and individuals who see an assessment as being undertaken on their behalf and with their endorsement and engagement. Examples might be village elders, land managers, agricultural cooperatives, or local or national governments. In practice, whether or not the members of the authorizing environment have provided formal authorization, the true test of whether the authorizing environment was sufficient is whether those stakeholders have a substantial ownership in the final products and a commitment to take actions based on the findings.

In some instances it may be appropriate to stimulate or encourage demand for an assessment. For example, in a situation where there is a lack of a consensus on aspects of the connection between ecosystems and people, an assessment might be proposed and communicated as being a useful tool for local decision makers to resolve particular issues, and thereby stimulate the demand for the assessment process and its outputs. In all cases, however, the decision to proceed with the assessment should be taken on the basis of actual rather than perceived demand, demonstrated through the recognition of, and approval as appropriate by, an authorizing environment. See Chapter 2 for further information on the importance and approaches for engaging users in the exploratory phase of the assessment.

Defining scope and boundaries

An assessment can be defined by its intended audience. If the primary users are national decision makers, then it will be a national assessment even though it might be examining ecological and economic processes from local to global scales (such as international trade in natural resources or climate change). If the primary users are international conventions, then the assessment is global. And if the primary users are a particular local community, then it is local.

The audience for an assessment is not the only factor involved in defining an assessment's scope. The scope ultimately depends on political, socioeconomic, and environmental circumstances that might constrain its boundaries. Even if the primary audience is a particular local community, ecological or social factors in the region might suggest that the scale should be larger than just one community. For example, if an issue of major concern to a community is water, then a river basin scale may ultimately be more appropriate since both ecological and social processes throughout that basin will strongly affect water availability in that community. The level of involvement of particular expertise or disciplines might constrain the themes addressed by an assessment; the level of involvement of users (or conflict between them) might also constrain the questions being asked. Constraints may arise as well from funding limitations, data shortages, or methodological constraints. In some cases these constraints can be overcome, such as through active recruitment of additional expertise. But in many cases the constraints will need to be acknowledged and considered during the design and process of the assessment.

1.3.2 Design stage

Once the needs and constraints have been identified, then the governance, content, and process for implementing the assessment can be determined. A thorough design phase, including consideration of funding and the ongoing engagement of users, is a key step in eventual success.

Governance

The governance (including leadership) of an assessment can be a critical factor in ensuring user engagement, raising funds, and overseeing progress in implementation of the assessment. It is also crucial to ensure legitimacy and credibility. A model that has been found to be effective in the MA and other assessments is having the assessment overseen by a technical Steering Committee or Assessment Panel and an associated "User" Advisory Committee or Assessment Board. In some cases, involving both the technical experts and users in a single committee might work well; in other instances, it may be more appropriate to establish a separate Advisory Group to represent the various users. The governance structure for the Millennium Ecosystem Assessment is provided in Figure 1.3.

Typical functions of a Steering Committee would be to:

- Promote coordination among the institutions and individuals carrying out the assessment;
- Develop the detailed assessment design (what information will be produced by which individuals and institutions);
- Increase the legitimacy of the assessment, and guard against bias from particular interest groups;
- Assure quality of assessment outputs;
- Design the outreach and communication activities; and
- Help to raise funds for the assessment.

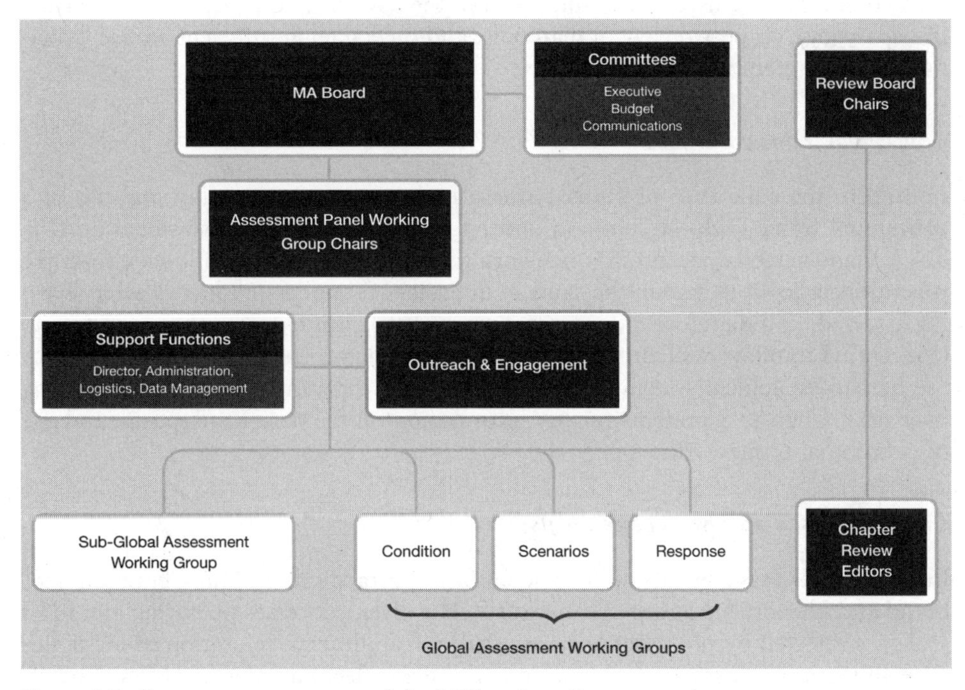

Figure 1.3. Governance structure of the Millennium Ecosystem Assessment.

The size and composition of the Steering Committee will of course vary, depending on the scope of an assessment. In the case of community efforts, it might consist of village leaders and researchers who will be involved in the assessment. In regional assessment activities, it is likely to include representatives of a number of different scientific networks and institutions within the region. Depending on the region or assessment, the Steering Committee could consist primarily of technical experts (in which case a separate advisory committee or Board should be established to ensure oversight of the process by users) or it could involve both a mix of technical experts and users. In either case, a small "executive" committee is likely to provide the most effective day-to-day oversight of activities.

If an advisory group is established, an important consideration is the extent to which that group should be strictly advisory or should have decision-making authority—for example, over the distribution of funding against priorities within the assessment. It is likely to be far more effective if the users hold decision-making authority either as part of the core steering group or as part of a formal board rather than serving in only an advisory capacity. If the intended audiences do not have some level of authority over the assessment, it is unlikely they will be sufficiently engaged in the process to fully use its findings. Moreover, without a formal requirement to respond to the needs of the audience, there remains a significant risk in any scientific assessment that the focus will begin to reflect the research interests of the scientists more than the needs of the decision makers.

Effective governance is also essential to ensure that assessment practitioners abide by the process of which they are part, including on agreements to share information and data used and resulting from the assessment (which helps to ensure transparency), on peer review of materials, and on engaging sufficient author expertise to ensure credibility and legitimacy.

Conceptual frameworks for assessment

Central to the coherence of an assessment is the design or adoption and use of a conceptual framework—a common understanding of what the assessment aims to do. A single agreed conceptual framework guides the assessment, allowing multiple practitioners to work within the same boundaries and understanding of what is being assessed, and therefore allows integration of the components of the assessment. Conceptual frameworks help clarify the complex relationships between elements of the human–ecological system, including how those relationships may be changing over time. Chapter 3 presents further information on the design, adoption, and use of conceptual frameworks.

Linking scales in ecosystem assessments

In many cases, assessments seek to identify important relationships between processes and phenomena at a particular scale. However, processes operating at a local scale are affected by processes at larger scales, from human migration to air pollution. In turn, processes operating at larger scales reflect cumulative or systemic effects of what is going on in a number of localities, such as land use and agricultural production. Assessing the condition and trends in ecosystems, their services, and human well-being requires an understanding of these cross-scale relationships. In many cases it will be possible to place an assessment in the context of others that have already been conducted at larger scales (and in all cases in a global context from the MA, the IPCC, or other global assessments). Important cross-scale linkages that might be considered as part of an assessment include:

- Linkages between environmental systems that operate at different scales;
- Linkages between institutional systems that play roles in ecosystem management and use; and
- Linkages with larger driving forces, including globalization, technological change, and institutional change.

The scale of the assessment will also influence the analytical units used for assessment and communication. In many cases there will be a geographic mismatch between the most appropriate units for assessment of ecosystems and their services (which depend on biophysical data and boundaries) and those for assessment of human well-being (which depend on sociopolitical and economic data and boundaries). See Chapter 4 for a further discussion on scale and units of analysis.

It is essential that an assessment (and thus the methods and data it uses) is at a scale that matches that of the biological and physical processes generating the ecosystem services being assessed and also at a scale relevant for the decision-making processes of the target audience. Sometimes these two scale demands can be reconciled by choosing one particular scale that satisfies both. Often this will not be possible, however, either because the scale gap between them is too large (which in itself is a warning signal that problems of management could arise) or because important processes are simultaneously occurring at several scales and interacting across them. In this case there is little choice but to do a multiscale assessment. In the case of cross-scale interactions, which tend to be common (for instance, national policy may determine a local outcome), it is far better to carry out an integrated multiscale analysis than simply to carve the assessment into a number of discrete scales that are independently assessed without trying to understand the linkages.

Integrated multiscale assessments are inherently more complicated and often more expensive than assessments at single scales. However, there are a number of ways to simplify a multiscale assessment. First, the assessment does not need to cover every possible scale. Usually there is enough flexibility in the scales of underlying processes that selecting about three key scales is sufficient: for instance, local, national and regional. Second, the sampling at a finer scale does not have to represent complete coverage at the coarser scale—it can simply cover what is determined to be a representative sample (this has been called a "sparsely nested hierarchy"). For example, the Southern African MA (SAfMA) assessed at the local scale in five areas, at the water basin level in the Gariep and Zambezi river basins, and at the regional level in Africa south of the equator. Third, limiting the number of issues to be assessed at multiple levels to a core set of perhaps three to five helps to keep the process manageable. Three ecosystem services were assessed by SAfMA, for instance, across all three scales: food, fresh water, and biodiversity. Finally, it is helpful if the indicators chosen at the various scales are either identical or bear some clear relationship to one another. It then becomes possible to trace how a particular issue is expressed at the various scales. More details about aggregating and disaggregating across scales are given in Chapter 4.

Bridging knowledge systems—linking formal and informal knowledge

Scientific assessments are based on a particular mainstream epistemology (way of knowing)—one that often pays little attention to informal (local, traditional, or indigenous—see Box 1.3) knowledge and that takes little in the way of cultural values into account. Scientists and policy makers alike have become aware that new assessment processes need to be robust enough to accommodate and value these different knowledge systems and the multiscale and multistakeholder nature of environmental concerns (Reid et al. 2006). A significant challenge for a multistakeholder assessment is to effectively bridge these traditional and scientific ways of knowing the world. A rich body of knowledge concerning the history of ecosystem change and appropriate responses exists within local and traditional knowledge systems. It makes little sense to exclude such knowledge just because it is not published. Moreover, incorporation of traditional and local knowledge can greatly strengthen the legitimacy of an assessment process in the eyes of many local communities. Finally, particularly at the scale of a local assessment, traditional knowledge may often be the primary source of historical information for the assessment.

It is helpful to think of information falling into one of four quadrants defined by the axes of formal–informal and tacit–explicit, because it makes clear what steps need to be taken to make such information usable in an assessment. Assessments fit into the formal–explicit quadrant (see Table 1.2). Information must either be made formal (by documenting it) or explicit (by placing it in the public domain), or both if it is to be used in an assessment. These techniques do not only apply to local and

Box 1.3. The semantics of different forms of knowledge

Informal (local, traditional, or indigenous) knowledge can add significantly to a scientific (formal knowledge) assessment, but the semantics that surrounds these different forms of knowledge can lead to confusion among assessment practitioners and users. The following working definitions are provided to help incorporate different knowledge systems into assessments and to bring some clarity and standardization to the terms used in the scientific and assessment literature.

- *Local knowledge* refers to place-based experiential knowledge systems. It may include traditional and indigenous knowledge; it is largely oral or practice-based and so is rarely documented.
- *Traditional knowledge* is the body of information, practices, and beliefs that evolves through adaptive processes and is handed down through generations through traditional transmission practices. If such knowledge relates to local ecology, then this is *traditional ecological knowledge*. Traditional knowledge is not necessarily indigenous, but it certainly has its roots in the past.
- *Indigenous knowledge* is the local knowledge held by indigenous peoples or the local knowledge unique to a particular culture or society. The term is usually only applied when referring to knowledge held by people who identify themselves as indigenous.
- The term *local and traditional knowledge* is the most encompassing of the informal knowledge systems. Although it may also incorporate elements of scientific or formal knowledge, it is often the most appropriate way of referring to a worldview that is different from mainstream science or government decision makers.

Table 1.2. A categorization of types of information.

	Tacit	Explicit
Formal	Private images or photos Unpublished models and databases Diaries	Ecosystem assessments Peer-reviewed papers, chapters, or books in the scientific literature Peer-reviewed databases
Informal	Opinions Experience Intuition Private beliefs and values	Oral traditional knowledge Indigenous knowledge, rules, and practices Communal beliefs and values Untested scientific databases

Tacit information is known only by individuals, whereas explicit information is shared, with some level of agreement.

Source: Fabricious et al. 2006.

traditional knowledge. For instance, much information is contained in the private experience of scientific experts and can be formalized and made explicit by questionnaires or interviews.

Capacity building

In the context of an assessment, capacity building is a continuous process aimed at strengthening or developing long-term relevant human resources, institutions, and organizational structures to carry out ecosystem assessments of relevance to decision makers and to act on the findings. Capacity building within an assessment has two objectives: to enhance the expertise of individual scientists to carry out ecosystem assessments and to enhance institutional expertise, particularly the science–policy interface, for effective adoption and use of the assessment findings.

Assessments may also provide important capacity-building opportunities through improving research capacities of universities and other research and training institutions, establishing baseline data for further assessments in the future, fostering an appreciation for scientific knowledge on the part of decision makers, and establishing or strengthening regional networks of experts.

Tangible ways to incorporate capacity building into the assessment implementation include developing a "fellows program" for young scientists to partner with more senior or experienced scientists engaged in the assessment, providing training courses on scenarios (or other assessment) methodologies, developing training materials, and forming partnerships with other institutions to expand the reach of these activities and to address decision makers' needs to build their own capacity to use the assessment's findings.

Evaluation

In order to learn constructively from assessments and to communicate those lessons to future assessment activities, a formal evaluation should be undertaken. Given the learning benefits both during the assessment period and afterward, two components of an evaluation may be considered: interim evaluations to allow for midcourse corrections during the assessment process and "post-hoc" evaluations after the initial phase to learn what worked, what did not, and where improvements could be made. Regardless of the specific mechanisms used, any rigorous and meaningful evaluation requires clear and early articulation of assessment goals, objectives, and Terms of Reference to provide a primary benchmark for evaluation and measurable indicators of success.

1.3.3 Implementing the assessment

Once an assessment has been designed around the requirements of identified users and the work plan has been developed to deliver the assessment, engage users, and communicate the process and findings, then the technical work of the assessment can begin. Of all the stages of the process, it is implementation that typically entails the greatest need for flexibility. In some cases the various components of technical work will need to be consecutive and sequential; in other cases, iterative and interactive. In all cases this will be determined by the constraints imposed on the assessment from finance or capacity limitations and by the options and opportunities available in terms of timing and resources.

The MA contained three basic components: assessment of condition and trends, scenarios, and responses. Although each of these can be undertaken by a separate working group of scientific experts, at many of the sub-global scales of the MA all three components were addressed by the same group. In many cases it would be better to undertake these three components sequentially, so as to first assess the condition and trends of ecosystems, their services, and human well-being; second, develop scenarios of change; and third, assess available and potential responses. In practice, however, these elements also need to interact closely, and what is learned from doing any one of these components can inform other technical parts of the assessment. The most important issue is that these components remain connected—through use of a joint conceptual framework, terminology, and approach. See also Chapter 5 for a discussion on the use of scenarios in the various stages of the assessment process, including as a tool to help determine the questions to be addressed by an assessment.

The condition and trends component should assess the priority ecosystem services and the associated drivers of change and impacts on human well-being that were selected during the design phase, based on the requirements and priorities determined from the user community and authorized by the assessment governance. Depending on data availability, it is preferable to consider the condition and geographical distribution and trends of the supply and demand for each service, as well as the effect of changes in ecosystems on their capacity to supply these services and the consequences of historical changes in these services for human well-being. A key element of the assessment of services will also be to consider the trade-offs that have been made between the supply of the various ecosystem services being assessed and, again, the consequences in these trade-offs for people.

The condition and trends assessment will by necessity focus on historical changes and impacts of ecosystem change on human well-being, and it should aim to use a sufficient variety of data, information, and indicators to be as comprehensive as possible regarding the selected ecosystem services. However, it is also feasible for this component to consider the foreseeable future and to thereby make a direct link into the scenarios assessment. See Chapter 4 for a full discussion on how to assess conditions and trends.

The scenarios component should aim to develop a set of scenarios providing descriptive storylines, supported by quantitative approaches, to illustrate the consequences of various plausible changes in drivers, ecosystems, their services, and human well-being. Scenarios are not attempts to forecast the future but rather are designed to provide decision makers with better understanding of the potential consequences of decisions. They also help to understand the uncertainty about the future in a creative way and can help explore new possibilities to respond to change. See Chapter 5 for details on how to use scenarios in an ecosystem assessment.

The responses component should aim to examine past and current actions taken to enhance the contribution of ecosystems to human well-being. In doing so, it should provide practical observations, tools, and guidelines for the various users on the effectiveness of actual and potential interventions, as well as promising options for action by a variety of stakeholders. Such interventions might include policies, practices, financial mechanisms, or communications and awareness-raising activities. See Chapter 6 for a full discussion on assessing the effectiveness of responses.

The peer review process is essential to ensure validation of the findings and to provide credibility to the process. Involving the user community in peer review of the findings also enables early feedback from the users on the utility of the assessment's outputs and is a key part of the communications strategy. The time involved to conduct a comprehensive review process—in terms of providing sufficient time both for reviewers to provide comments and for the assessment team(s) to incorporate (or justify the exclusion of) comments can be considerable. However, given the importance of credibility, in all cases the value of peer review will outweigh the time and costs of the process. In addition, it can be extremely helpful for increasing the transparency and objectivity of the assessment process if responses to review comments are made publicly available.

Assessments will need to develop review processes that are tailored to their own circumstances and to the scale and the context in which it is undertaken. In general, however, review processes should aim to meet the following criteria:

- The review process should be independent. An independent party not involved in the governance or operations of the assessment must have the authority to determine whether reviewer inputs have been sufficient and whether the comments have been handled adequately.
- Relevant governments (for the scale at which the assessment is conducted), NGOs, regional institutions, and other organizations as appropriate should be contacted in advance to identify appropriate reviewers, and reviews should be requested from all these sectors.
- Reviewers should be requested with the aim of obtaining a balanced representation of views within the scope of the assessment and among scientific, technical, and socioeconomic perspectives.
- Reviewers should include experts involved in the larger- and smaller-scale assessments within which the assessment is nested or that are contained within the assessment.

- All written review comments, and the responses to those comments, should be archived and made publicly available.

1.3.4 Communicating ecosystem assessments

Assessments can succeed or fail depending on the communications strategy, and one of the most important decisions made by the assessment team or advisory body will be how to distribute limited resources between technical work and the communications component. Both the process and the outputs of the assessment are critical to communications, and the impact of the assessment will depend as much on communicating the legitimate and credible process as it will on communicating the policy-relevant findings.

The primary purpose of an assessment is to meet the needs of decision makers at the scale at which the assessment is conducted. In doing so assessments will almost always produce products, including written reports, audiovisual materials, and other products tailored to the needs of the decision makers at that scale. The specific nature of these products will depend on the circumstances and scale involved and on the specific needs of the users. Thus, for example, a regional (multicountry) assessment would likely produce a technical report volume and a summary for decision makers aimed at the needs of any regional governing bodies and the national policy makers of countries within the region. At the same time, it could also produce products that could empower local communities or contribute to educational activities. On the other hand, a local community assessment may produce only a single report, since the users of the assessment might largely be the same as the people producing the assessment. Documents published as components of an assessment should adhere to a set of criteria for the preparation, peer review, and approval of the documents approved by the assessment governing body.

1.4 How to link an assessment to the decision-making process

Section's take-home messages

- Entry points for incorporating the results of an ecosystem services assessment into decision-making processes occur at all levels of governance and are important for both development officials and those approaching problems from an environmental perspective.
- Opportunities for mainstreaming ecosystem services can be categorized into four intersecting entry points: national and subnational policies, economic and fiscal incentives, sector policies, and governance.
- The information that will be most effective in eliciting constructive responses—and hence the most appropriate focus of an assessment process—depends on the place of a particular issue in the policy life cycle of environmental issues.
- Four key elements ensure that an assessment is policy relevant: identifying the questions for policy or decision making that the assessment should try to answer; answering those questions; stating the levels of certainty (or uncertainty) associated with the findings; and presenting the answers in a separate Summary document, an Executive Summary, or a "Main Messages" section.

1.4.1 Understanding the decision-making context

Decision makers as diverse as mayors, national economists, natural resource managers, and conservation planners can use an ecosystem services assessment to explore the links between ecosystems and economic development, gaining a better understanding of how their goals both affect and depend on ecosystems and their services. Those working in the social and economic development community often base their analysis on elements of human well-being such as health or food; through the participation in, or outputs of, the assessment, they can connect those to ecosystem services. The environmental conservation community, on the other hand, can start from an environmental perspective and assess the implications of conservation actions on development and human well-being.

Entry points for incorporating the results of an ecosystem services assessment into decision-making processes occur at all levels of governance and are important for both development officials and those approaching problems from an environmental perspective. Many entry points are found at the national or provincial level. Some, such as the Millennium Development Goals or international trade and investment, are at the global level, but these usually have more detailed counterparts at the national or local level. Opportunities for mainstreaming ecosystem services can be helpfully categorized into four intersecting entry points: national and subnational policies, economic and fiscal incentives, sector policies, and governance (see also Table 1.3).

- *National and subnational policies:* The preparation of national and subnational trade, economic growth, or immigration policies provides important entry points for managing the cumulative demand and impacts on ecosystem services from individual or multiple sectors. Ministries of the environment, treasury, development, and planning, among others, may play a role.
- *Economic and fiscal incentives:* Fiscal measures such as subsidies, taxes, and pricing influence decisions throughout the economy, from firms and farms to factories and households. They can be designed to create incentives to sustain and efficiently use ecosystem services, as well as to create disincentives for activities that drive ecosystem degradation.
- *Sector policies:* Ministries of commerce and industry, science and technology, and agriculture and forestry, among others, can play an effective role in advancing policies and actions that sustain ecosystem services. Environment agencies can work with other government agencies and departments to develop information, tools, and analyses that help make the connection between ecosystem services and the attainment of sector goals.
- *Governance:* Strong governance is at the heart of sustaining ecosystem services. This includes public participation in decisions that affect or depend on ecosystem services, a free press, and requirements to provide information to the public, including regular indicators of ecosystem health. All branches of government also have a role in providing oversight. Such mechanisms enable citizens to hold governments and businesses accountable for their use and management of ecosystems.

Table 1.3. Entry points for mainstreaming ecosystem services

Entry points	Ministry, agency, or organization	Examples of decision processes
National and subnational policies and plans	Development & planning Environment Treasury Physical planning, emergency planning, and response	Poverty reduction strategies, land use planning, water supply, and sanitation Protected area creation, climate adaptation strategies National budgets, public expenditure reviews, audits Integrated ecosystem management of coasts, river basins, forest landscapes, and watersheds
Economic and fiscal incentives	Finance	Subsidies, tax credits, payments for ecosystem services, import duties, and tariffs Tax policies to support easements or promote alternative energy technology, pricing regulations for water
Sector policies and plans	Commerce and industry Science and technology Agriculture Forestry Environment/ natural resources	Corporate codes of conduct/standards, assessment of new technologies Applied research, technology transfer, business capacity building Extension services, best management practices Forest sector action programs, mapping initiatives, concession management State of the environment reports, strategic environment assessments, environmental impact assessments, information/tools, legal instruments
Governance	Prime minister's or president's office, justice ministries, legislature, local government bodies	Decentralization policies, free press, civil society, accountability of government through elections, access to information and decisions, judicial review, performance indicators

The examples provided for each entry point are not intended to be exhaustive but rather to illustrate the variety of ways that ecosystem service considerations can be incorporated into decision processes.

Source: WRI 2008.

Decision making is complex and multidimensional and occurs at all scales, from individual to international. Decision makers can be individuals or groups of people, in private or public sectors. Assessments take place in the context of this political economy. Thus, asking the right questions to inform decision making requires close interaction with the decision makers who are the target audience of the particular assessment. It also requires sufficient knowledge of the decision-making process to be able to tailor the assessment, and particularly the communications from the assessment, appropriately. The effectiveness of an assessment in terms of influencing policy and actions can also be constrained by how open and willing the political arena is to the concept of bringing science to bear on decisions. Better science can certainly lead to better decision making, but it does not guarantee it.

Decision makers have at least three strategies to choose among in responding to information presented in an ecosystem assessment:

- Do nothing.
- Compensate people who are affected by ecosystem change.
- Prevent (or reduce) ecosystem degradation through policy and regulations, market-based instruments (taxes and/or subsidies), other means of social control, or some combination of these approaches.

Even with overwhelming evidence of the impacts of ecosystem change on people, considerable pressure may be required before decision makers take action. The information that will be most effective in eliciting constructive responses—and hence the most appropriate focus of an assessment process—depends on the place of a particular issue in the policy life cycle of environmental issues (see Figure 1.4):

- *Stage 1*: Identification of a particular environmental or development issue by a small group of "pioneers" but no broader awareness either by society at large or by decision makers.
- *Stage 2*: Lobbying by "action groups," often a denial of effects by some groups of stakeholders, and incipient awareness but no action by decision makers.

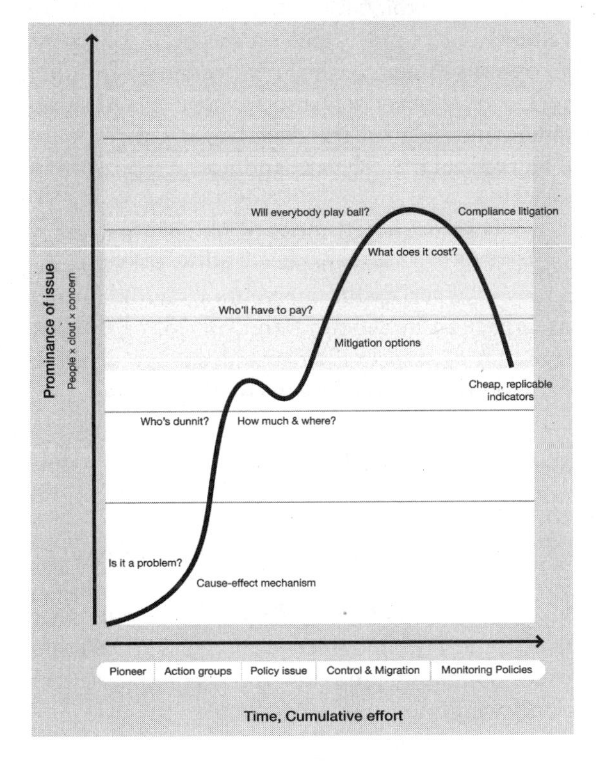

Figure 1.4. The environmental issues "life cycle," showing how interest and public perception of environmental issues changes through time.
Source: Tomich et al. 2004

- *Stage 3*: Widening acceptance of the potential or actual environmental or development issue, with mounting awareness and pressure from civil society for decision makers to take action.
- *Stage 4*: Debate on evidence for understanding and causes of the issue, which has now become a policy issue.
- *Stage 5*: Inventory and assessment of prevention and mitigation options and their environmental, socioeconomic, and administrative costs and benefits.
- *Stage 6*: Negotiations on prevention or mitigation of impacts.
- *Stage 7*: Implementation, monitoring, and enforcement of prevention or mitigation actions.

Although the details will differ between various environmental and development issues, the course of events in local, national, and regional issues could follow similar paths, at least if actions somehow respond to both the numbers of people concerned and the intensity of their concern. For an issue in Stages 2 and 3 of this cycle, an assessment would be most effective if it focused on testing the validity of Stage 1 "suspicions" about a link between an undesirable ecosystem change and an impact of human well-being. Establishing a probable cause-and-effect chain as opposed to "mere coincidence" or "spurious correlation" is important at this point as a basis for sound policy intervention. This information also could help build broader support for action and undermine resistance from vested interests. In these stages there also is a need to estimate the likely magnitude of impacts (are the effects big or small?) since initial uncertainty may range over several orders of magnitude.

Once awareness and support is formed for action on a particular issue, the debate may shift focus to understanding the causality of the problem. Perceived gaps in the quantification of impacts or in a causal explanation of the phenomenon are major obstacles in Stage 4. Various stakeholders may agree on the need for an inventory of prevention and mitigation options (Stage 5), or the process may be more adversarial, with each group applying evidence selectively and advocating a position serving its own interests. An assessment focusing on the effectiveness of response options and scenarios will therefore be particularly relevant and useful for issues in Stage 5 and 6.

Beyond stage 6, the information needs tend toward those of monitoring rather than assessment, although it is likely that a future assessment will allow for more informed adaptive management and a review of successful interventions in the future, and some problems that have apparently been understood and solved, or at least brought under control, may re-emerge in a new cycle if situations change or if the initial understanding proves to be incorrect or responses ineffective.

1.4.2 The importance of being policy relevant: moving beyond scientific reviews

A scientific assessment is fundamentally different from a scientific research project or review. An assessment leads to a different product and has a different audience and communications strategy. It is not a research initiative and does not generally seek to undertake new research, but rather applies the judgment of experts to existing knowledge generated from the scientific community (and other forms of knowledge) to provide credible answers to policy-relevant questions.

The audience of an assessment is decision makers, typically in government (national and local), business, or civil society, and particularly those operating at the scale of, and within the scope of, issues addressed in the assessment—although there may also be other key audiences. For example, an assessment of water resources and associated ecosystem services in a particular river basin will be of direct interest to water and land mangers in the river basin, but it also might generate useful insight for managers in adjacent and other basins and for decision makers in other sectors, such as health professionals interested in water-related health issues.

Four key elements enhance the policy relevance of an assessment: identifying the questions for policy or decision making that the assessment should try to answer; answering those questions; stating the levels of certainty (or uncertainty) associated with the findings; and presenting the answers in a separate Summary document, an Executive Summary, or a "Main Messages" section.

First, it is important to identify the relevant questions for policy or decision making that the assessment should try to answer. A "policy-relevant" question is one that is asked by a user group, audience, or decision maker or one where the answer to the question can be used to justify or support a decision or action that directly or indirectly affects the allocation of public or private resources (money, time, land use rights, etc.). A question is also policy relevant if it informs conceptual understanding of how the world works or otherwise informs decision makers about the context for their decision making and is linked to the decision-making process. The most effective way to identify policy-relevant questions is to survey users directly to determine their "user needs." Given the scope of a particular assessment, it might also be useful to identify other policy-relevant questions that the assessment could answer even if they have not yet been formally posed by users (i.e., other policy-relevant issues may emerge from the assessment that are not yet on the agenda of the assessment users but might be useful to consider or be aware of). Chapter 2 provides further guidance on ensuring that an assessment is responding to user needs.

Second, to be policy relevant, the assessment will need to answer the questions identified. Certainly it will need to be designed around answering the user needs. But also it will often make sense to structure the assessment outputs around the policy-relevant questions that were identified in the planning stages. This is a different approach than the one common in the scientific community, which would allow for a more comprehensive analysis of the underlying science. Such a "science-led" approach runs the risk of losing the policy-relevant assessment outputs among the presentation of other materials.

A critical issue that each assessment must confront is whether or not to provide specific recommendations for action. In general, experience shows that scientific assessments are far more effective if they do not make specific recommendations for policy or other actions. The experience of global assessments to date is that it is most effective to present findings that are policy relevant, as described here, but not policy prescriptive. The rationale for this approach is that the choice among policy options is never a strictly scientific one. Science can legitimately inform policy options, but actually selecting the most desirable option is a political role. The transition in an assessment to selecting preferred options can undermine the credibility and legitimacy of the science itself.

While the approach of providing policy-relevant but not policy-prescriptive findings is by far the best approach for global, regional, and even national assessments, it may not apply in all cases. For example, community assessments will involve as experts most of the individuals in certain communities (since they will have local knowledge of the historical changes in the system, which can complement the scientific information being contributed from outside scientists.) In such a situation, the experts are also the users of the assessment. It thus would become somewhat artificial for that assessment to draw a boundary between the presentation of policy-relevant findings and the selection of the most promising options.

The following checklist of guidelines around presentation and content can help ensure that key questions are appropriately addressed by an assessment:

Presentation

- Present and synthesize information that is needed to answer the question: no more and no less.
- If an authoritative source for information already exists from other assessments at the same or other scales, then cite it directly (e.g., IPCC, World Water Development Report, Global Biodiversity Assessment). In this way new information can be added to the existing assessment base, rather than being re-created.
- Do not just list the existing data or information—provide the assessment team's judgment concerning best estimate or range, likelihood, probabilities, uncertainties, and reliability if the quantity or value is relevant to policy decisions (i.e., rate of forest loss, rate of local species extinction, economic contribution of pollinators to agricultural production, etc.).
- Present all credible points of view or scientific results and identify areas of scientific disagreement. A "credible point of view" is one present in the peer-reviewed literature unless there is a body of evidence that clearly negates it. Different perspectives should be included not only because they might be right, but also because by not citing different viewpoints or providing judgment on them directly, you open the assessment up to charges of being biased and not rigorous, thereby affecting the credibility of the findings.
- Do not make specific policy recommendations. However, it is perfectly feasible to explore the likely consequences of different options using an "If . . . then . . . " approach—so, for example, a policy-relevant, nonprescriptive finding might be that "If fertilizer use is reduced by 20% in a particular landscape, then freshwater quality in the area is very likely to be improved, leading to measurable increases in local fish harvests."

Content

- Generally, research findings should not be presented for the first time in the assessment. All information in an assessment should have already been subject to a peer review process (or equivalent for nonformal knowledge systems—see Annex 2 for guidance on procedures for incorporating nonpublished, nonpeer-reviewed material into assessments). However, it is fine to use existing peer-reviewed models with different datasets—so, for example, a scenarios exercise might run new estimates of changes in drivers through existing models.
- New datasets generated through assessments should undergo peer review as part of the assessment process.
- If data or information do not exist to answer a policy-relevant question that

has been identified in user consultations, do not just ignore the question—conclusions such as "there is insufficient scientific information to determine whether . . ." are still relevant to decision makers.

The third way to enhance the policy relevance of the assessment is to state the levels of certainty (or uncertainty) associated with the assessment findings. The treatment of scientific uncertainty within assessments is also a particularly important factor influencing the credibility of the process. Clear statements of what is unknown are often as influential for policy makers as are statements of what is known with relative certainty. This is particularly important in the main messages at the highest level of reporting, such as in any Executive Summary. The assessment of the state of knowledge should reflect both the type and amount of evidence (e.g., observations, interpretation of model results, or expert judgment) and the level of peer acceptance or consensus. Further information on the treatment of uncertainty can be found in Chapters 3, 4, and 5.

Any agreed language for stating levels of certainty should be appropriately and consistently used. Some summaries may not need to use them at all or may use them only once or twice. Specific uncertainty language is most appropriately used for statements of most direct relevance to policy decisions. So the statement "alkaline lakes are characterized by lower species richness than freshwater lakes" does not need a statement of certainty, for example. The statement "climate change will cause the most significant changes to ecosystem services in the following ecosystems in the region: xxx and xxx (*medium certainty*)" should be accompanied by a statement of certainty, however, based on the evidence found in the scientific literature. Uncertainty language should also be used when the finding reported depends on a judgment of the group of experts who wrote the chapter rather than a reporting of "fact." So, for instance, "although the magnitude of the positive feedback to global warming due to changes in each individual ecosystem feedbacks has *low certainty*, the balance of evidence indicates a *high certainty* that together these ecosystem changes have contributed to the recent increases in Arctic air temperature." However, there is rarely a need to include uncertainty language when reporting on trends or data unless the interpretation of the data required a judgment call by the experts. So "the rate of forest loss over the past 20 years in the district has been 4 ha/year" does not require a statement of certainty, although typically the level of certainty would be reflected when a range is presented—that is, "152 hectares" implies a different level of certainty than "100–200 hectares."

The final way to improve the policy relevance of the assessment is to present the answers to the questions identified at the outset in a Summary document, an Executive Summary, or a "Main Messages" section. Summaries can contain more than this, but they should be sure to contain the material most relevant to policy makers. A good summary

- Highlights key policy-relevant findings supported by arguments/evidence;
- Frames the information presented in relation to the questions asked of the assessment by users—this often follows the assessment outline and structure but does not need to have key findings for all the sections;
- Is brief but not cryptic and avoids ambiguity;

- Has a tone consistent with that of the underlying assessment;
- States the degree of certainty when necessary, highlighting robust findings and key policy-relevant uncertainties;
- Contains no literature references but can have internal references to the underlying assessment;
- Identifies gaps in knowledge and so is clear about what is known and what is not known; and
- Is reader-friendly, written in an accessible and not too technical style (see Box 1.4).

Box 1.4. Some useful writing suggestions for assessment reports

These suggestions are based on comments received during the MA peer review process.

- Avoid writing assessments to read like textbooks aimed at giving a general appreciation for the systems discussed and possible problems they may be experiencing. Actual assessment information about problems, threats, and actionable items becomes buried in these sections. Generally speaking, putting the educational material first is a turnoff for decision makers. The problems and actions should be discussed first. Necessary educational material should be intermingled with that discussion. General descriptions of systems could be appended or referenced to other available information. The assessment is what needs to be up front, not the background.
- Avoid passive voice and focus on definable measures and actions. For example, "there are reasons to believe some trends can be slowed or even reversed" is a totally ignorable statement in policy circles. If there are some opportunities for reversal, state precisely what we believe they are, as best we know.
- Statements like ". . . might have enormous ramifications for health and productivity . . . ," while they seem to the scientist to be strong because of the word "enormous" are actually politically impotent because of the word "might." If data were used in the assessment, what does they say about what "is" happening? What can we recommend, based on best knowledge, about what actions would be effective?
- Statements like "There is a long history of concern over the environmental effects of fishing in coastal habitats, but the vast scope of ecological degradation is only recently becoming apparent (citation)" is a case where something strong could be said, but it is weakened by putting the emphasis on the late arrival of this information and knowledge "becoming apparent." It does not matter so much when the degradation was discovered, what matters is that it was. Cite the source and say "fishing practices are causing widespread destruction."
- Do not use value-laden, flowery, or colloquial language (e.g., "sleeping dragon," "elephant in the room," etc.).
- Statements like "we do not yet have clear guidelines for achieving responsible, effective management of natural resources" could result in a legitimate policy response of "OK, so we'll wait until we do." Instead, the statement could be changed to recommend what needs to be done, such as "if clear guidelines were developed, then . . . ".

Additional Resources

Chapter 6: Assessment Process, in Millennium Ecosystem Assessment, *Ecosystems and Human Well-being, Volume 4: Multiscale Assessments* (Washington, DC: Island Press, 2005). This chapter of the Millennium Ecosystem Assessment sub-global working group provides an analysis of the lessons learned in implementing the sub-global assessments of the MA. Available from www.MAweb.org.

Ecosystem Services: A Guide for Decision Makers, (Washington, DC: World Resources Institute, 2008). This guide demonstrates how a city mayor, a local planning commission member, a provincial governor, an international development agency official, or a national minister of finance, energy, water, or environment can use information from an integrated ecosystem assessment to strengthen decisions. It forms a companion volume to this practitioner's manual. Available from www.wri.org/publication/esa.

MA Sub-global assessment network. The network of sub-global assessments established during or since the MA provides a forum for exchange of ideas and best practices. Representatives from the various assessments meet periodically to share experiences. (www.MAweb.org)

The Global Environmental Assessment Project is an international, interdisciplinary effort directed at understanding the role of organized efforts to bring scientific information to bear in shaping social responses to large-scale environmental change. The focus of the Project was the growing number of assessments—ranging from the periodic reports of the Intergovernmental Panel on Climate Change to the Global Biodiversity Assessment to the MA—that had recently been conducted in support of international policy making. Its focus was to understand the impacts of environmental assessments on large-scale interactions between nature and society and how changes in the conduct of those assessments could alter their impacts. The Project attempted to advance a common understanding of what it might mean to say that one effort to mobilize scientific information is more "effective" than another. It tried to view such issues from the perspectives of the scientific experts involved in producing assessments, the decision makers at multiple scales who use those assessments, and the societies affected by the assessments. (www.hks.harvard.edu/gea/geadescr.htm.

ANNEX 1

A step-by-step guide to the assessment process

Exploratory Stage *(steps to be taken simultaneously)*

1. Establish the need or demand for an assessment of ecosystem services and the consequences of ecosystem change for human well-being at a particular scale.
2. Convene a technical and user planning group to determine the feasibility, scope, and extent of the assessment. Draft a scoping study.
3. Interact extensively with the intended users of the assessment—for instance, recruit an advisory board consisting of end users and stakeholders and modify the scoping document based on their subsequent review.
4. Explore funding support and generate funding proposals.

Design Stage

5. Establish the governance and institutional structure: technical executive group, advisory board, contracts and memoranda of understanding between participants and funding agencies.
6. Communicate the objectives, agenda, and opportunities for involvement to the stakeholders through Internet, newsletters, and presentations.
7. Assemble an initial assessment team based on competence in the topic areas selected, experience in the geographical area, availability to participate, and credibility with stakeholders.
8. Through an assessment team workshop, draft an Implementation Plan detailing who is responsible for what actions, by when. Review it in the light of comments from the advisory board.
9. Agree on a shared conceptual framework among all assessment practitioners and users.

Assessment Implementation and Communications Stage

10. Perform preliminary assessments of each of the focus areas identified in the scoping study, focusing on the three elements of the assessment: condition and trends, scenarios, and responses.
11. Continue interactions with the intended users of the assessment through workshops and briefings. This will help ensure the assessment is still of value and aid in development of a communications strategy that considers how best to deliver the preliminary findings and assessment reports to different user groups.
12. Circulate the assessments among the assessment team and affiliated technical experts for "internal" peer review.
13. Key issues for which there is no credible (preferably published, peer reviewed) supporting material may require a focused, high priority research action. Commission or encourage such studies as early as possible and submit the results to a peer-reviewed, open literature publication process.
14. Revise the preliminary assessments in the light of the review comments. Draft an

Executive Summary (typically 2–5% of the length of the full text) to accompany the full assessment.

15. Circulate the updated draft to at least one independent technical expert per topic area.
16. Revise the draft in the light of the expert reviewers' comments. This is the last opportunity to incorporate new material.
17. Finalize communications and outreach strategy and begin implementation.
18. Circulate the updated assessment draft to the experts who previously reviewed it, additional experts, the users and other stakeholders identified during scoping, and the advisory board.
19. Revise the draft in the light of the comments received to create the penultimate draft assessment report. Document the revisions made or the reasons for not making them.
20. Submit the penultimate draft to the advisory board for approval. This board or a separate review board should ensure that due diligence has been performed in addressing reviewer comments.
21. Incorporate any revisions required by the board and perform final formatting and graphics editing. No substantive additions or deletions not mandated by the board should be introduced at this stage.
22. Publish and distribute the assessment through appropriate media: in a special issue of a journal or as a stand-alone volume. Publication on the Internet is recommended as a secondary source.
23. Extract and distribute communication products from the assessment, taking care not to deviate from the spirit of the accepted text. These may include policy summaries, graphic-rich brochures, media releases, posters, maps, CD-ROMs, videos, and radio interviews.

Scoping for Subsequent Assessment

24. Convene a workshop consisting of technical participants, stakeholders, users, and donors to analyze the strengths and weaknesses of the assessment and to document the lessons learned.
25. Explore the possibilities for and scope of a follow-up assessment.
26. Find an institutional home for the records and archives of the project and as a point of contact until such time as a possible future assessment.

ANNEX 2

Guidance for using nonpublished/ nonpeer-reviewed sources

Guidance for using nonpublished and nonpeer-reviewed sources into the MA was provided to all authors involved in the process. The following text is taken from this guidance and may serve as a useful reference for assessment initiatives that are seeking to benefit from the incorporation of knowledge from sources beyond the peer-reviewed scientific literature.

Because considerable materials relevant to MA Reports, in particular, information based on indigenous, traditional, or local knowledge or information about the experience and practice of the private sector, are found in sources that have not been published or peer-reviewed (e.g., industry journals, internal organizational publications, nonpeer-reviewed reports or working papers of research institutions, proceedings of workshops, personal communication, etc.) the following additional procedures are provided. These have been designed to make all references used in MA Reports easily accessible and to ensure that the MA process remains open and transparent.

1. Responsibilities of Coordinating, Lead and Contributing Authors

Authors who wish to include information from a nonpublished/nonpeer-reviewed source are requested to:

a. Critically assess any source that they wish to include. This option may be used for instance to obtain case study materials from private sector sources for assessment of adaptation and mitigation options. Each chapter team should review the quality and validity of each source before incorporating results from the source into an MA Report.
b. Send the following materials to the Working Group Co-Chairs who are coordinating the Report:
    - One copy of each unpublished source to be used in the MA Report
    - The following information for each source:
        - Title
        - Author(s)
        - Name of journal or other publication in which it appears, if applicable
        - Information on the availability of underlying data to the public
        - English-language executive summary or abstract, if the source is written in a non-English language
        - Names and contact information for 1–2 people who can be contacted for more information about the source.
c. Information based on personal communication from individuals with indigenous, traditional, or local knowledge, or direct input as a member of a working group by an individual with indigenous, traditional, or local knowledge should be handled in the following way:
    i. In situations such as local assessments where extensive use of local and traditional knowledge will be involved, the assessment must establish a process

of validation for the findings as part of the application by the assessment to become a component of the MA. The features of such a validation process are described in Section 4.2.

ii. Metadata concerning the personal communication (e.g., names of people interviewed, dates and types of notes recorded, presence or absence of self-critical review notes by the researcher, sources of 'triangulation', etc.) should be made available to the Co-Chairs of the Working Group.

iii. Where an individual provides direct input of indigenous, traditional, or local knowledge as a member of a working group, the individual should provide the Working Group Co-Chairs coordinating the report the following information:
  - Basis for knowledge of the particular issue (length of time living in the area, individuals from whom historical information was obtained, etc.)
  - Names and contact information for 1–2 people who can be contacted for more information about the source.

2. Responsibilities of the Review Editors

The Review Editors will ensure that these sources are selected and used in a consistent manner across the Report.

3. Responsibilities of the Working Group Co-Chairs

The Working Group Co-Chairs coordinating the Report will (a) collect and index the sources received from authors, as well as the accompanying information received about each source and (b) send copies of unpublished sources to reviewers who request them during the review process.

4. Responsibilities of the MA Secretariat

The MA Secretariat will (a) store the complete sets of indexed, nonpublished sources for each MA Report not prepared by a working group and (b) send copies of nonpublished sources to reviewers who request them.

5. Treatment in MA Reports

Nonpeer-reviewed sources will be listed in the reference sections of MA Reports. These will be integrated with references to the peer-reviewed sources stating how the material can be accessed, but will be followed by a statement that they are not published.

References

Fabricious, C., R. Scholes, and G. Cundill. 2006. Mobilizing knowledge for integrated ecosystem assessments. In *Bridging scales and knowledge systems: Concepts and applications in ecosystem assessment*, ed. W. V. Reid, F. Berkes, T. Wilbanks, and D. Capistrano, 165–82. Washington, DC: Island Press, for World Resources Institute.

MA (Millennium Ecosystem Assessment). 2005. *Ecosystems and human well-being: Current state and trends*. Washington, DC: Island Press.

Reid, W. V., F. Berkes, T. Wilbanks, and D. Capistrano, eds. 2006. *Bridging scales and knowledge systems: Concepts and applications in ecosystem assessment*. Washington, DC: Island Press, for World Resources Institute.

Tomich, T. P., K. Chomitz, H. Francisco, A. M. N. Izac, D. Murdiyarso, B. D. Ratner, D. E. Thomas, and M. van Noordwijk. 2004. Policy analysis and environmental problems at different scales: Asking the right questions. *Agriculture, Ecosystems and Environment* 104 (1): 5–18.

WRI (World Resources Institute). 2008. *Ecosystem services: A guide for decision makers*. Washington, DC: WRI.

2

Stakeholder Participation, Governance, Communication, and Outreach

Nicolas Lucas, Ciara Raudsepp-Hearne, and Hernán Blanco

What is this chapter about?

In this chapter the case is made for developing an assessment process that is relevant, credible, and legitimate to stakeholders and end users, and guidance is offered on how to implement these principles. The assessment process is as important as the reports it produces, and therefore it is crucial to be strategic when managing stakeholder engagement and participation, organizing the assessment process, and communicating with stakeholders. Section 2.1 describes the principles of relevance, credibility, and legitimacy and offers suggestions on how to achieve these principles within the assessment process. Section 2.2 presents approaches for exploring the need for an assessment and then initiating the process. Sections 2.3, 2.4, and 2.5 provide a blueprint for organizing and managing the assessment process, based on past experiences with integrated ecosystem assessments at the sub-global scale. And, finally, section 2.6 offers information on how to develop and communicate messages based on assessment findings.

2.1 How to ensure relevance, credibility, and legitimacy

Section's take-home messages

- The three fundamental qualities of a sound assessment are relevance, credibility, and legitimacy. Is assessment information significant in relation to an actor's priorities or decision-making issues? Does the assessment meet standards of scientific rigor and technical adequacy? And do participants perceive the assessment process as unbiased and meeting standards of political and procedural fairness?
- Assessments need to strike a balance between relevance, credibility, and legitimacy and, if possible, promote synergies between them. At the local level, these qualities are sometimes at odds with those at the national or global level.
- The core values of relevance, credibility, and legitimacy are best achieved through strategic and effective participation in the assessment process.

How an assessment process is organized is just as important as the information it produces. A sound assessment design helps ensure that audiences will be interested in the assessment findings and able to make use of the information in their decision making. Normally, scientists, managers, and policy makers focus their efforts on ensuring the credibility of technical information by attempting to produce authoritative, believable, trusted technical reports. However, recent research and practice has shown that underemphasizing the relevance and legitimacy of ecosystem information, which depend on how an ecosystem assessment is conducted, can lead to ineffective and ultimately futile exercises (see Box 2.1) (Mitchell et al. 2006, Farrell and Jager 2005, Eckley 2001).

Information-gathering processes are more likely to be seen as legitimate and to be used more constructively in policy making when they are transparent and involve those who will be affected by any decisions influenced by them. This section explains the three fundamental qualities of a sound assessment—relevance, credibility, and legitimacy—and gives examples of how they might be achieved.

2.1.1 Relevance, credibility, and legitimacy

Relevance, or salience, refers to the significance of assessment information in relation to an actor's priorities or decision-making issues (Cash et al. 2002). Relevance is measured both in terms of contents (the questions that the assessment answers, the issues it addresses) and timing (information that feeds too late or too early into a decision process is likely to be considered irrelevant). Issues of interest to researchers may not be the same ones that interest decision makers, and, ultimately, information will be relevant if judged useful by decision makers rather than by researchers, as this is the goal of an assessment.

Assessment information must help decision makers address relatively precise questions. For example, if the policy goal is to reduce rural poverty, the information must answer what ecosystem services are critical to rural livelihoods, what is their status of degradation or enhancement and how they might change in the future, what are the actual and expected consequences for human well-being, and how different decisions or policies will modify the supply of ecosystem services to the poor.

Designing an assessment in order to produce relevant information begins with framing the assessment around questions of importance to target audiences (see

Box 2.1. Lessons from international processes

In 1995 some 1,500 scientists released the *Global Biodiversity Assessment*. This 1,152-page high-quality report was produced largely independently from its primary intended audience—namely, governments who were parties to the Convention on Biological Diversity. As a result of this legitimacy deficit, the information was not welcome and ultimately the parties ignored the report for the most part (Raustiala and Victor 1996). Similarly, the early stages of the Intergovernmental Panel on Climate Change (IPCC) lacked legitimacy in the eyes of participants from developing countries, who saw that greater participation of their scientists was needed (Agrawala 1998). This was addressed in later stages, and the IPCC in time became enormously influential globally.

Source: Philippines and Chile SGA teams, personal communication.

section 2.4). A "user needs assessment" can be conducted to find out what these questions are. Both the assessment design and audience needs may evolve over time, and it is therefore important to have regular communication with targeted audiences throughout the assessment to maintain a fit between information and needs. Finally, how the information is conveyed to different audiences has implications for how relevant they find it (see section 2.6).

Temporal relevance is also important. If information is being developed to fit directly into a decision-making process that will take place during a certain time period, the assessment must be concluded before that period. This may mean making decisions to drop elements of the assessment in order to meet the deadline, as a slightly more compact assessment delivered on time is more useful than a longer assessment report that is obsolete. Firm dates for when the assessment findings will be released and ready for use should be negotiated early on in the process.

Credibility refers to whether the assessment meets standards of scientific rigor and technical adequacy. Sources of knowledge must be considered trustworthy and/or believable, along with the facts, theories, and causal explanations invoked by these sources (Cash et al. 2002). Technical information intended to feed into decision making is always subject to criticism, especially from those who stand to lose from the resulting decisions, so it is important that the assessment process builds the credibility of the information generated (Ranganathan et al. 2008).

What is judged to be believable will differ between communities, and therefore it is important to understand key audiences well (see section 2.6). While some audiences may consider published scientific papers to be a credible source of information, some local actors may find information to be more credible if it fits into their cultural context and knowledge system. Local and traditional knowledge can be included in a scientific assessment, but it must be validated and reviewed in a manner that is credible to all stakeholders. Assessments are powerful policy tools because they aim to include information and expertise from multiple disciplines, and sometimes also different forms of knowledge. Integrating these different sources of information results in more broadly credible findings.

The process by which knowledge and information are collected, sorted, and made sense of needs to be highly transparent in order to achieve broad credibility. Before and during the assessment process, transparency can be emphasized by developing and publicizing process documents that describe:

- The assessment process;
- The conceptual framework;
- Guidelines for participation;
- Guidelines for how to validate unpublished knowledge, including local and traditional knowledge;
- Guidelines for dealing with uncertainty; and
- Guidelines for the review process.

Legitimacy refers to whether an actor perceives the assessment process as unbiased and meeting standards of political and procedural fairness, and whether the process considers appropriate values, concerns, and perspectives of different actors. Audiences judge legitimacy based on who participates and who does not, the process for making those choices, and how information is produced, vetted, and disseminated

(Cash et al. 2002). Legitimacy refers to the politics of the information-gathering process. It is a function of the political context into which the information is released. In essence, the goal of a strategy to establish legitimacy is to involve users to the point where they invest sufficiently in the process that they will adopt and use the information produced, or at least not reject it. This requires keeping users informed and giving them a chance to influence the process so that, at the very least, they do not feel threatened by it. And in an ideal situation, they "own" the process and its results. The extent of user engagement can range from reporting on progress to information users to the users being centrally involved in the production of the information. In any event, at the very least users need to be comfortable with the motivation, origins, and general orientation of the technical work.

During the exploratory and design stages, the global Millennium Ecosystem Assessment (MA) proponents spent a lot of time discussing the project with the targeted stakeholders to build the legitimacy base. Meetings were held with leading countries and figures within the three main ecosystem-related international conventions (the Convention on Biological Diversity, the Convention to Combat Desertification, and the Ramsar Convention on Wetlands), and presentations made to their governing bodies, which, early on in the MA process, formally expressed their interest and expectations. In addition, multiple contacts were established with 22 representative institutions that were eventually invited to form the Board of Directors for the project. Throughout the assessment, efforts were made to open the process to the participation of national and local stakeholders through a variety of means (from user forums to the circulation of drafts for review). In the end, this huge effort delivered a very sound legitimization, as well as a very valuable platform for outreach.

Many of the MA sub-global assessments observed that navigating the politics of stakeholder participation often results in a trade-off between advancing with technical work and ensuring the legitimacy of the process. Broadening and deepening participation may conspire against expediency in technical tasks and also subject scientists and experts to policy discussions that they are not prepared for or comfortable with. A balance between technical work and participation must be found without compromising either. Stakeholders can sense when their participation is merely lip service or considered a "distraction from the real work." Even minimal stakeholder participation can be made more effective by paying attention to the following:

- Ensure participation is built into actually relevant stages of the assessment process.
- Invite key stakeholders to participate in the assessment's governance structure.
- Establish communication channels between stakeholders and technical experts involved in the assessment to clarify uncertainties and verify assumptions.
- When stakeholders are invited to contribute based on their experience, make sure they can recognize their inputs in the analysis and reports, and inform participants that their contribution and participation will be properly acknowledged in outputs.
- Where possible, ensure stakeholder inputs are recorded and made available to them.
- When the assessment is meant to serve the needs of a large organization, ensure

that the assessment is formally recognized and supported by the governing bodies of the organization.

2.1.2 Synergies and trade-offs between relevance, credibility, and legitimacy

Relevance, credibility, and legitimacy are interconnected, positively and negatively. In the previous section they were presented separately for the sake of clarity, but in reality actions to strengthen one may have both positive and negative consequences on the others (Ranganathan et al. 2008). For example, the involvement of government officials in the review process may be undertaken initially to ensure the relevance of the drafts being produced, but it is likely to enhance the legitimacy for the assessment and add to its technical credibility with some audiences as well.

On the other hand, the enhancement of one quality can come at the expense of another. Traditionally, the scientific community has favored credibility over legitimacy and relevance as its main concern when generating information, the assumption being that technical and scientific information needed to be walled off from the potentially biasing influence of politics, prepared with quality control of peer review, and delivered to decision makers in the form of reports (Cash et al. 2002). This strategy does protect the technical credibility of the information, but often at the expense of making it less relevant to policy makers. Similarly, when policy makers are allowed to develop the substantive contents of the assessment reports (as opposed to only deciding on the questions that frame it), the legitimacy of the reports may in some cases be enhanced, at the cost of becoming less technically credible.

A good design for an assessment needs to strike a balance and, where possible, promote synergies between relevance, credibility, and legitimacy. This balance needs to be found across scales as well. The experience of many sub-global assessments was that local relevance, credibility, and legitimacy were sometimes at odds with achieving these qualities at national or global scales. In these cases, the assessment team might choose to focus on meeting the expectations of the primary audience, but they might also develop strategies to work with audiences at other scales. For example, an extensive and inclusive review process is a strategy for achieving credibility and legitimacy across scales. Most of the MA sub-global reports were reviewed by global experts in addition to experts and stakeholders from the most relevant scales.

2.1.3 Participation as a strategy for achieving relevance, credibility, and legitimacy

The core values of relevance, credibility, and legitimacy are best achieved through strategic and effective participation in the assessment process. To participate is to share, and a participatory process involves having different stakeholders engaged in an interactive process that promotes knowledge and information exchange and that allows them to express their positions and interests on issues.

An integrated ecosystem assessment requires blending knowledge and perspectives from many different points of view. It usually aims to influence audiences with different interests and information needs. In order to maximize impact, it is essential that a wide range of actors participate throughout the process, either as contributors or audiences—or both. This helps identify key issues that matter most within a given context, strengthens the analysis of perceived changes in ecological and social systems, and builds ownership of the assessment's findings among audiences who are supposed to follow up with action. Participation can take the following forms:

- Being consulted on the need for an assessment;
- Being consulted on key questions framing the assessment;
- Receiving information about assessment progress, findings, and opportunities to participate;
- Contributing knowledge to the assessment report;
- Contributing contextual information about an ecological or social system;
- Being consulted on the condition and trends of ecosystem services and human well-being in a region (practitioners and holders of local knowledge);
- Attending public hearings about assessment process and findings;
- Attending education or capacity-building workshops on assessment process and findings;
- Participating in the assessment process as student interns or fellows of the assessment;
- Becoming a member of the advisory committee;
- Being a formal end user of the assessment products;
- Reviewing assessment materials; and
- Acting as a partner for the dissemination of assessment findings.

Some forms of participation are considered necessary for an assessment (e.g., consulting potential end users on key questions to frame the assessment), while other forms may or may not be necessary or possible (e.g., consultations on the conceptual framework or methodology to use, or parliamentary intervention to obtain formal endorsement). The assessment team generally decides on participation strategies based on the desired impact of the assessment on different stakeholder groups. The team will generally emphasize spending energy on encouraging the most important stakeholder groups to take ownership of the process and outcomes. Communication with target stakeholders should be sustained throughout the entire assessment process.

Improving information through participation

In addition to building relevance, credibility, and legitimacy, encouraging broad participation in the writing of the assessment report can improve the quality of the findings. To assess ecosystem services in relation to human well-being, technical teams must first of all be multidisciplinary and incorporate understanding of natural and social sciences. (See Chapter 4.) At smaller scales, the incorporation of local stakeholder knowledge into integrated assessment has recently been established as critical (Pahl-Wostl 2003). This is because local stakeholders have particular and unique knowledge about the ecosystems where they live and work, as well as about their own associated well-being. Often the links between ecosystem services and human well-being are complex and obscured and must be teased out with contextual knowledge. In addition, published data about ecosystems and societies in many parts of the world is scarce and can be much enhanced with local knowledge. In the Peruvian sub-global assessment, communities gathered to assess the condition and trends of water and soil by consensus in an area with little data availability.

The MA encouraged the use of local, traditional, and practitioner knowledge in assessments. However, in order to be credible and useful to decision makers, the MA Conceptual Framework stated that "all sources of information, whether scientific, traditional, or practitioner knowledge, must be critically assessed and validated as part of the assessment process through procedures relevant to the form of knowledge" (MA 2003). Chapter 4 describes methods for validating different types of knowledge.

Beyond sound information: Participation in assessments as a moral issue

To build relevance, credibility, and legitimacy, an assessment process needs to be inclusive and open to different stakeholders. Openness, transparency, and participation are therefore justified on practical grounds—that is, the need to ensure that assessment information is effectively used by decision makers. But the social dimension of an assessment is not fully captured by this justification. Assessments, especially when they involve governments or publicly held information, and precisely because they are intended to influence decisions with a public impact, are part of the cultural life of a community and the advancement of science in it. The right to participate in such a process has been recognized by the Universal Declaration of Human Rights in Article 27 (1): "Everyone has the right freely to participate in the cultural life of the community, to enjoy the arts and to share in scientific advancement and its benefits." Moreover, an open assessment is a process of democratization of both information and governance.

In practice, the composition of stakeholder groups invited to join the assessment process is under control of those proposing the assessment. An assessment of relevant stakeholders and their information needs is one way to establish the target audiences (see section 2.3), but keeping the process open to interested parties promotes participation as a right, even when the degree to which all stakeholders can participate will be limited by time, funding, and language constraints.

2.2 How to establish the need for and scope of an assessment—the exploratory stage

---

Section's take-home messages

- It is of strategic importance to develop the specific objectives of an assessment according to what is possible—in order to keep expectations realistic—and what is needed and sensible—in order to get as much support as possible from potential stakeholders. The subtle, early work of an assessment involves discussing its rationale with a diversity of stakeholders, which requires understanding and integrating their diverse perspectives and interests.
- The first and most obvious scope to be defined is the geographic extent or spatial scale of the assessment. There are trade-offs to consider for both large- and small-scale efforts.
- In the exploratory phase, a user needs assessment can explore who should be included

in the assessment, and in what capacity. This would identify the information that different stakeholders need about ecosystem services. Ideally it would also produce a database listing all potential stakeholder groups, plus information about their relation to specific ecosystem services and their potential and capacity for engaging in the assessment.

An ecosystem assessment is a technical, social, and political process. It is technical as it entails the use of sound science, it is social as it engages a diversity of actors in a collective endeavor, and it is political as it involves different stakeholder interests. An assessment may be mandated by law, but it still requires leaders who first sense the need for an assessment and then work with others to make it happen. Assessment initiators might be direct users of ecosystem services, scientists, decision makers, or any other actor interested in the environment and its link to human well-being. This section aims to help these early assessment promoters; it also highlights some of the main challenges that might be encountered in the early stages of the process and suggests ways to deal with them.

2.2.1 How to define the need for an assessment

Sometimes the need for an assessment will be evident to users or decision makers. Sufficient and valid conditions to consider initiating an assessment process include:

- A lack of information about trends in ecosystem services;
- Particular events (such as unusual floods) that highlight the lack of knowledge about ecosystems in a region;
- Observed decreases in the quality of ecosystem services;
- Latent or overt conflicts among users over the use of ecosystem services;
- Poor or unsatisfactory levels of governance of natural resources;
- Impacts on human well-being; and
- Negative or uncertain prospects for the future of local ecosystems.

These conditions are common in most developing and industrial countries. At this early point in the assessment, it is of strategic importance to develop the specific objectives of the work according to what is *possible*—in order to keep expectations realistic—and what is *needed* and *sensible*—in order to get as much support as possible from potential stakeholders. Promoters of an assessment are responsible for the subtle, early work of discussing the rationale for an assessment with a diversity of stakeholders, which requires understanding and integrating their diverse perspectives and interests. A number of MA sub-global assessments held exploratory workshops, for example, to discuss the need for an ecosystem assessment with local people.

The promoters of an assessment then have to exercise the art of balancing expectations and possible results of the assessment. In particular, promoters will have to prepare and present the project in such a way that the principal stakeholders—users of the ecosystem and users of the assessment results—can find value in it. (See Boxes 2.2 and 2.3.) This is a challenging task, particularly considering that the results of such an assessment are, in general, medium to long term and not as tangible as most typical development projects. The task is made easier if the promoters are themselves stakeholders in the system and well integrated into the local decision-making context. In many cases, however, promoters are scientists or nongovernmental

Box 2.2. Initiating an ecosystem assessment

Defining the need for an assessment in the Laguna Lake Basin, Philippines

Prior to the Philippines MA subglobal assessment, the scientific community of Los Baños was heavily involved in different research programs that focused on the Laguna Lake Basin and its management, because of its location next to important urban centers. The assessment was viewed by the University of the Philippines Los Baños (UPLB) as an opportunity to connect to the ongoing efforts of many agencies for collaborative management of the lake. The assessment initiators (one forestry scientist and one social scientist at UPLB) then invited key people from the Laguna Lake Development Authority (LLDA), the agency with the mandate to manage the region, to form the core assessment team. With this group formed, other agencies engaged in the research, use, and management of economic activities in the lake basin were invited to participate. Agencies like the Philippine Marine and Aquatic Resources Management and Development of the Department of Science and Technology and the Bureau of Fisheries and Aquatic Resources of the Department of Agriculture sent their own scientists to the workshops and meetings. Together with the core team, a scientific committee was loosely organized and contributed to the assessment of the different ecosystem services.

To validate the objectives of the assessment, the core assessment team met with LLDA executives, who facilitated a meeting with the Secretary of the Department of Environment and Natural Resources and with government executives of the two main provinces involved. This was when the need for such an assessment was fully established and legitimized, which ensured the acceptability of the process to various stakeholders. The proposed assessment objectives and process were finally presented to a wider forum in Manila, where representatives of other sectors were invited in order to get inputs from the wider audience. In retrospect, the assessment team felt that it should have included voices from the local communities in the region.

Recognizing the potential of assessment work in the Salar de Atacama, Chile

In the Salar de Atacama, Chile, a lack of access to information on water quality and quantity in one of the driest areas in the world made it difficult for the users of that resource to design an acceptable water management plan. Latent conflict existed among these resource users (including mining companies, tour operators, and the indigenous local communities), in part due to the lack of information. Because the MA was soliciting assessment projects and offering some funding, capacity building, and a credible international network, researchers in Chile developed a proposal for a subglobal assessment. They saw the assessment as an opportunity to bring stakeholders together to discuss different points of view and to establish credible baseline information on water resources that were at the center of stakeholder conflict. Indigenous people were a central stakeholder in this assessment.

Source: Maria Victoria O. Espaldon, University of the Philippines Los Banos, personal communication.

organizations (NGOs) who must work closely with local leaders in order for them to eventually take ownership of the process and outcomes.

Assessment promoters may also be faced with unsupportive contexts. For instance, local authorities might have a record of negative experiences with environmental research projects that have been extractive as opposed to collaborative. If this is the case, promoters will have to make every effort to differentiate the assessment

Box 2.3. Getting an assessment started in Western China

In October 2006, the Ministry of Science and Technology of China funded a follow-up assessment of China at the national level, based on the experience of the MA subglobal assessment of Western China, the results of which are currently being used to inform policy and action in that area. The municipality of Qing-Yang was selected as one of the case studies for the China assessment. In November 2007 the assessment team organized a workshop in Qing-Yang, which was attended by the mayor of the municipality and officials from several public divisions.

After the workshop, the mayor expressed a strong interest in exploring trends in local ecosystem services, and future scenarios for the region. The Qing-Yang municipality asked the national assessment team to help them develop tools for ecological management and to develop a decision-support system for development and conservation of the Dong-Zhi Tableland, which is the economic and political center of Qing-Yang Province and includes four counties.

In 2008, Qing-Yang municipality collected data and maps and provided them to the national assessment team for analysis. At the same time, the Qing-Yang municipality organized six officials and 70 technicians from different divisions to investigate trends in ecosystems and driving forces. These officials and technicians were trained by scientists from the national assessment team in methods of data collection and processing. The setting up of data collection procedures and training has established conditions that will be amenable to future ecological management and planning by the local government.

Source: Prof. Dr. Tian Xiang YUE, Institute of Geographical Sciences and Natural Resources Research, Chinese Academy of Sciences, personal communication.

process from previous research endeavors. Spending enough time explaining how and why the assessment can be useful to local stakeholders, and discussing the perceived and potential links between ecosystem services and local health, well-being, and economic development, is key. One great value of an assessment process is the chance to raise awareness about human dependence on ecosystem services. An assessment process can contribute to developing knowledge about the state of ecosystems so that decision makers can make smart and sustainable decisions to benefit human well-being. When referring to outcomes, promoters can thus emphasize the following benefits of conducting an assessment:

- The generation of and access to relevant information on the condition and trends of ecosystem services that support local populations in different ways;
- The opportunity for constructive stakeholder participation and collective learning opportunities;
- The opportunity for directed capacity-building processes;
- The integration and coordination of diverse private and public initiatives; and
- State-of-the-art information for better ecosystem management and increased human well-being.

2.2.2 How to define the scope of an assessment

The most obvious scope to be defined is the geographic extent or spatial scale of the assessment. Sub-global assessments within the MA demonstrated that an assessment approach centered on the MA conceptual framework can be implemented from very local scales to large, regional scales. The reasons (or motivations) behind the need for an assessment will dictate the appropriate scale. For example, if assessment stakeholders are interested primarily in how land use is affecting water quality in the region, the watershed scale may be the appropriate one to consider. The choice for the preferred spatial scale is also determined by the ecosystems themselves, how they function, and the location of the ecosystem users.

The MA advocates a multiscale assessment process, which integrates assessments at diverse scales into one overarching assessment. (See Chapters 3 and 4.) The benefits of conducting an assessment at multiple scales are many (see MA 2003); however, this type of process will inevitably require more time and resources.

There are trade-offs to consider when selecting the appropriate spatial scale. A large scale might allow for a comprehensive analysis of relevant social and ecological systems and interactions between them. But it will demand more resources, and the findings may be more difficult to integrate into a focused policy-making process. Larger scales will also pose a number of challenges to the participatory process, the governance structure, and the communication and outreach initiatives. For example, stakeholders might not recognize the relevance (or urgency) of the assessment if it is not entirely focused on issues of importance to their local context.

A small-scale assessment, on the other hand, might facilitate coordination and integration with the policy- and decision-making process, but it might lose a wider perspective on important issues occurring outside the chosen scale. A smaller-scale process might also facilitate the organization of the participation, governance, communication, and outreach aspects of the assessment, although this is not always the case. Because assessments at smaller scales are focused on issues of direct relevance to local stakeholders, they tend to be more interested in the outcomes, as well as more divided on issues that are controversial, and the participatory process can thus become an extremely intensive part of the assessment process.

In practical terms, there is usually a demand generated for an assessment at a specific scale, generally coming from the most enthusiastic stakeholders. They will be able to determine (sometimes with the help of scientists) what spatial scale is most relevant to their decision making and to the ecosystem services of interest to them. When seeking to define the boundaries of the assessment area, there are two aspects to consider:

- *The natural (geographic) boundaries*: For example, river basins are natural boundaries of ecosystems; they also allow for an integrated analysis of the relationship between water resources, soil, flora, and fauna. The exception might be the case of groundwater resources (and related ecosystems) that in certain cases do not follow the superficial boundaries of basins. Additionally, some situations might not follow the rationale of geographic boundaries; one such case would be when environmental services are imported from a different country or region.
- *The political–administrative boundaries*: Most administrative divisions within countries (and between them) do not follow natural limits such as river basins or ecosystems. Institutions, governments, and decision-making processes, which often have a large influence on the way ecosystems are managed, are crucial to the assessment process and hence are a logical way to define spatial scale.

Other factors that need to be determined when deciding on the scope of the assessment include how many and what ecosystem services will be assessed, over what timeframe the ecosystem services and human well-being will be assessed, and whether sophisticated technical approaches such as modeling will be used. These are all dictated by the goal of the assessment, and while the initiators will have an idea in advance about these factors, they will need to be discussed further as the governance structure and technical team are assembled (see section 2.3). Chapter 4 goes into further detail about how to develop the substantive elements of the assessment.

2.2.3 How to identify and engage relevant and diverse actors

A user needs assessment at the beginning of the assessment process is a good way to begin a stakeholder engagement strategy. This will include the communities, institutions, organizations, groups, and individuals that may be interested in the assessment process and findings, either because they affect ecosystem services or because they themselves are affected by changes in such services. A user needs assessment would ideally produce a database listing all potential stakeholder groups, plus information about their relation to specific ecosystem services and their potential and capacity for engaging in the assessment. More important, the needs assessment would include a survey to identify the information that different stakeholders need about ecosystem services.

This database could be added to and improved on as the assessment progresses. One way to initiate the user needs assessment is to have a social scientist, preferably with field experience, conduct desk research and then visit the diverse communities, organizations, and individual users. After several interviews, the social scientist will be able to start consolidating a users database. However, there are many other ways to conduct a user needs assessment, depending on the context and scale of the assessment.

A systematic understanding of the main ecosystem and assessment users, as well as the main social, economic, and political components of the system, is central to the success of the assessment process. This may sound like common sense, but a thorough assessment of user needs is not often done. This exercise is not only about getting access to information but, most important, about getting to know the relevant stakeholders and initiating or strengthening relationships with them. It is through discussing a potential assessment with stakeholder groups that a vision of the assessment is formed and the more engaged stakeholders get on board formally.

In addition to the focus on users, a database identifying relevant initiatives related to ecosystem services (present and future) is also useful. This would include initiatives such as new legislation, conservation movements, or health projects that the assessment process might influence or contribute to. The database of potential users and relevant initiatives can be complemented with a simple background study of the region that would set the stage for developing the assessment. This might include:

- A local history;
- Important economic activities and basic human well-being indicators;
- Recent issues and conflicts and the way they have been dealt with;

- Channels of communication used for communication amongst stakeholders; and
- The identification of leaders (formal and informal).

A brief report synthesizing the user needs assessment and assessment context study, and highlighting the implications for the assessment process, will define how the assessment process is developed. For instance, when it comes to communicating and involving stakeholders, knowledge about channels of communication and key local leaders will be crucial. In and of itself, the user needs assessment—if done respectfully—will contribute to building a relationship with users.

Once the user needs assessment is completed, further work is required to strengthen the involvement of stakeholders. As mentioned in section 2.2.1, the exploratory stage requires the patient work of meeting with a diversity of relevant actors, getting to know them, communicating the essentials of the assessment process, and listening to their interests and positions. The following issues will need to be discussed with stakeholders in order to define their roles within the developing assessment process:

- The spatial scale of interest;
- Important ecosystem services;
- Availability of information about ecosystem services and human well-being;
- Governance structure for the assessment;
- Participation and communication activities;
- Coordination with existing initiatives;
- Consideration of traditional knowledge; and
- Possible uses of the assessment results.

At this stage, the assessment initiators will have to decide who the key users are. (See Box 2.4.) These might be the groups that are most affected by changes in ecosystem services and/or groups that have the most influence over managing the services. In some cases, it may be the groups that are most interested in the process, regardless of their relationship with ecosystem services. It may be useful to formalize the relationship with these users by producing a memorandum of agreement or a formal definition of user roles. Section 2.3.2 describes the process of organizing a governance structure that will include key stakeholders. Despite this formalization, the process can remain open to all additional interested stakeholders. The role of each stakeholder may be different, however, and will need to be defined (see section 2.1.3 for a list of possible roles of stakeholders). The resulting group of key stakeholders who now have a formal relationship to the assessment will be those who are then convened to formally launch the assessment process (see section 2.4.1 on convening assessment participants).

Promoters of an assessment will have to be cautious about participants' expectations. Public participation experiences tend to fail because of early unrealistic expectations that are not managed in a timely manner and are eventually not met. The best way to manage expectations and keep them realistic is by making ongoing efforts to be extremely clear and honest in terms of what an assessment can and cannot achieve and about what is uncertain. In more practical terms, though this exploratory stage will require bilateral meetings, it is recommended that multiactor

Box 2.4. User groups in the MA subglobal assessments

Almost all subglobal assessments involved national, regional, or local government agencies as users. A large number of them identified local communities, NGOs, universities, and research institutes as important users in addition to the government agencies. The private sector (for example, the tourism industry, mining companies, and logging companies) was involved in only five assessments, despite the MA goal to support a greater role for the private sector in environmental decisions. This may be because the assessment teams did not have experience in working with the private sector. One lesson from this experience is to dedicate some time or team member to working on a relationship with important private-sector stakeholders.

Indigenous communities were involved in six assessments. Assessments that conducted user needs assessment during the exploratory stage of the assessment found them to be very valuable. Assessments also found it useful to have team members with good networking skills in order to involve key decision makers who might make use of assessment findings.

meetings be carried out after a first round of meetings. These tend to be an easier way to check unrealistic expectations, as all actors are confronted with one another's interests and positions. In addition, written agreements and terms of reference for all activities should be produced, clearly establishing what the assessment process does and does not intend to do.

2.3 How to organize an assessment process

Section's take-home messages

- The first step in organizing an assessment process is to identify what resources, particularly expertise and financing, are needed. The roles within an assessment team that need to be filled include scientists, communication specialists, a facilitator, local experts with contextual knowledge, social and political analysts, and a project coordinator.
- The organizational structure of an assessment includes a governance structure that will keep the assessment work on track and ensure the process is credible, legitimate, and relevant, along with work teams established around different themes or stages of the assessment.
- One way to organize a governance structure is to convene an advisory group—up to 20 representatives from diverse stakeholder groups. The assessment may need a separate technical advisory group competent in specific subjects that are the focus of the assessment.

2.3.1 How to identify the resources needed for an assessment

Who to include in an assessment team

Promoters of an assessment are not necessarily the people best prepared to coordinate and carry out all the assessment work. An assessment process might require a range of capacities not found within one organization. In this exploratory stage, one crucial task is therefore to define the capacities that will be required and to identify the possible assessment team. Critical criteria to consider when consolidating the assessment team include the following:

- *Capacities* to deal with the relevant assessment components (ecosystem goods and services, human well-being, conditions and trends, scenarios, responses, communication and participation).
- *Credibility of the work team.* Respected scientists and practitioners will add weight to the assessment findings.
- *Politically relevant organizations/individuals* that may help ensure that the process and results are effectively considered and integrated into decision making.
- *Local organizations/individuals* as part of the team to ensure legitimacy and relevance. Local experts (practitioners, leaders, or simply individuals with good knowledge of the history of the area) can also provide contextual information and knowledge that is critical to developing assessment findings.
- *Ethnic/cultural balance*, to be coherent with the reality in the assessment area. For instance, in an area with a significant indigenous population, the project team might include relevant indigenous organizations or individuals, either directly in the project team or indirectly in the assessment governance structure.
- *Gender balance.* The legitimacy of the process is strengthened by a balanced gender ratio.

The assessment team is often a partnership of organizations from diverse sectors. Organizations from the public and private sectors might increase the likelihood that the results will be considered and implemented; scientific organizations (e.g., universities) might have access to relevant information and capacities; NGOs might have networks and links to key users. A partnership of diverse organizations may make the assessment process (and its results) more transparent and accountable and will facilitate communication and outreach activities. On the other hand, partnerships will add complexity to the governance structure. Therefore a clear and transparent way of working is required, including a structure defining responsibilities and persons in charge.

In some cases, the capacities required for the assessment are not available in the project area, and the assessment team may be brought in from a different area. In these cases, care should be taken to discuss the reasons for this with local stakeholders and seek their support. And it will probably be necessary to integrate local people with diverse capacities and knowledge into the assessment team in order to contribute knowledge about local trends and relationships between ecosystem services and human well-being.

The concrete capacities required by an assessment team will vary according to the ecosystem services being considered and the human well-being indicators of interest. In general, there will be a need for professionals from the natural and social sciences. Ideally, the assessment team should also include people with expertise in participation, communication, and outreach.

The following is a short list of roles to fill within an assessment team:

- Scientists and/or practitioners with expertise in measuring the condition and trends in the ecosystem services of interest and relevant human well-being indicators;

- Communication and outreach specialists;
- An experienced facilitator for managing participatory processes;
- Local experts with contextual knowledge of trends in ecosystem services and human well-being;
- Geographic Information System (GIS) specialists for mapping conditions and trends;
- Social and political analysts with local knowledge and experience to work on relevant responses;
- A project coordinator to manage the process; and
- Designated persons to manage the governance body and review process.

How to estimate the necessary funding

The MA experience shows that the provision of seed funding was an appropriate mechanism for facilitating the generation of assessment processes; however, this type of funding is not always available. In fact, MA sub-global assessments found it difficult to obtain outside funding, as assessments were not seen as activities generating immediate "results."

One important result of the exploratory stage is the estimation of the whole project budget. Among the aspects that will define the extent of funds needed are the spatial scale of the assessment; the size and nature of the technical effort (e.g., the specific ecosystem services to be assessed and the way they will be studied); the size and nature of the participatory, communication, and outreach processes; and the availability of information and local capacities. Budgets will vary widely depending on all these details and therefore it is impossible to offer more concrete guidance on the size of budget needed. MA sub-global assessment budgets ranged from $15,000 to several million dollars (see Table 2.1). For many of the sub-global assessments, in-kind contributions were a significant way to add needed resources.

Despite the fact that core funding might come from one specific source (e.g., international donor or a central government), it is highly desirable that local potential donors—if available—be approached and invited to contribute. This will not only add resources but, more important, might provide an opportunity to gain the trust and commitment of relevant stakeholders. For instance, private companies that directly benefit from ecosystem services (e.g., forestry, mining, or fisheries) might be good candidates to contribute project funding, provided the funds do not affect the outcome or legitimacy of the process. Companies' dependence and impact on ecosystem services can be assessed through tools such as the "Ecosystem Services Review" recently developed by the World Resources Institute. Assessment promoters can approach local potential donors at the beginning of the process and learn about their interests and needs. Their inclusion in a user needs assessment will facilitate fundraising.

The participatory, governance, communications, and outreach activities within the assessment might require a nonnegligible proportion of the total budget. In the MA, project leaders often did not budget significantly for these kinds of activities, which were considered to be "add-ons" at the end of the core work. Yet as already mentioned, the usefulness of the assessment hinges on a successful and well-planned participatory process, which thus needs to be budgeted from the beginning.

Table 2.1. Budget for the San Pedro de Atacama Subglobal Assessment, Chile

N.	*Item*	MA	In-kind mining companies	In-kind public agencies	Other donors	Total (US$)
1.	Salaries and consultants	35,000				35,000
2.	Meetings, workshop and travel expenses	10,000	5,000	3,000	5,000	23,000
3.	Basic information		5,000	5,000		10,000
4.	Materials and products	5,000			10,000	15,000
5.	GIS and satellite imaging		5,000			5,000
6.	Tourism assessment requirements			2,000	10,000	12,000
7.	Water assessment requirements	5,000	10,000			15,000
8.	Partners	6,000				6,000
9.	Administration costs	3,000				3,000
10.	Overhead	6,000				6,000
	TOTAL (US$)	**70,000**	25,000	10,000	25,000	**130,000**

2.3.2 How to design a governance structure

The ultimate goal of setting up a governance structure for the assessment is to ensure the relevance, credibility, and legitimacy of the process and findings, as described earlier. This means ecosystem (and assessment) users must, to some extent, own the process. If this is not achieved, the assessment will become an academic exercise with little real impact.

One way to organize a governance structure is to convene an advisory group, a set of representatives from diverse stakeholder groups associated with the assessment (see section 2.2.3). Advisory groups were used in a number of sub-global assessments in different capacities. (See Box 2.5.) To be a manageable forum for discussion and decision making, ideally it should include a limited number of persons (fewer than 20) who represent the community at large.

There are no fixed rules on how an advisory group should function. In terms of power to make decisions, its role may range from being solely advisory (decisions are made by the project team) to having final decision-making responsibility (with decisions executed by the project team). In general, the objective of the advisory group is to accompany the project development, providing information about user needs and advice to the project team. This ensures the relevance of the ultimate assessment outcomes. The advisory group is also a key mechanism through which important stakeholders (some of whom are key decision makers) can learn to trust the credibility of the assessment work and take ownership of the process and outcomes. Building the relevance, credibility, and legitimacy of the process and

Box 2.5. User engagement and governance structure in the Southern African assessment

The Southern African Millennium Ecosystem Assessment (SAfMA) was designed and implemented in a way that encouraged the participation of multiple stakeholders and users of the assessment information. SAfMA set out to be user-driven, and stakeholders played an important role in its governance. At the regional scale, an Advisory Committee (AC) with 10 members of different groups was responsible for representing the interests of the different stakeholders, balancing the various interests within the region, and creating a receptive policy environment for the assessment's outputs. The AC directed the work of the teams conducting the assessment and endorsed SAfMA outputs at each stage of the process. At the other scales of assessment, User Advisory Groups played this role. In this way, users had ownership of the process, and they endorsed and signed off on outputs that they considered to be credible. The AC also ensured that the different assessment teams were adhering to agreed-upon schedules and timelines, as this was imperative for the integration of the assessment findings across scales. The AC played a role as well in steering the project through difficult phases and provided leadership and guidance.

Due to its multiscale nature, SAfMA stakeholders were varied. The different categories of SAfMA users were engaged in a variety of ways. In addition to their being in the AC and User Advisory Groups, other stakeholders were appointed to review panels and were involved in intensive meetings and workshops at the various scales. Stakeholders were also engaged through a SAfMA Fellowship Programme, where individuals from stakeholder organizations were invited to become SAfMA Fellows. This involved participating in SAfMA activities, reviewing SAfMA documents, and assisting with outreach and dissemination of SAfMA materials. SAfMA Fellows also acted as bridges between SAfMA and other programs in the region.

Users expressed their needs in the meetings and workshops held. Prior to the start of the assessment, the need for information had been stressed at numerous national workshops, in various State of the Environment Reports, and in Strategy Documents of the Southern African Development Community (a regional grouping of Southern African countries). The needs of users were also ascertained through direct consultation in workshops and meetings and through the participation of user groups in the review of various reports and documents.

outcomes ensures that the assessment is considered and integrated into targeted decision-making processes.

The role of the advisory group can be both political (as described above) and technical. The assessment process may require a technical advisory group competent in specific subjects that are the focus of the assessment. Such a group can act as a sounding board to deal with complex and contentious issues. The social and ecological complexity of the assessment questions, as well as the spatial scale of the assessment, will determine the need for a more or less technical advisory group. In either case, the advisory group will play a role in managing the review, ensuring a balanced and fair assessment process, and making decisions on how to present contentious results.

Once a decision has been made in terms of setting up an advisory group, it needs to be formally implemented. Ideally, during the user needs assessment (see section

2.2.3) promoters will have already identified key organizations and individuals that could be part of an advisory group. The rights and responsibilities of the advisory group, as well as the working style, are usually defined in writing as terms of reference (ToR). The very first meeting of the advisory group should be devoted to discussing and approving the ToR. The ToR can define:

- Whether members are invited as at-large members (e.g., as individuals) or as institutional representatives;
- How final decisions will be negotiated and who will have the final say;
- How conflicts will be resolved;
- Whether there is financial compensation involved for advisory group work;
- How many meetings must be attended; and
- The structure of the advisory group.

The group's structure will depend on the size and the scope of the assessment. Larger assessments may require members to assume oversight of particular aspects of the assessment process, such as the budget, the review process, or the communication process. Advisory groups will almost always require a chairperson.

The decision to include political leaders in the advisory group is not straightforward. Political leaders can raise the political profile of the assessment, but they can also subvert the credibility and legitimacy of the process in the eyes of some stakeholders. It may be more effective to involve technical representatives of the political leaders, ensuring that the leaders are informed about the assessment activities and invited to participate in some of them.

2.3.3 How to organize the work team

During the exploratory stage, assessment team members are proposed and discussed (see section 2.3.1). Once the work team has been assembled, dividing the assessment work and formalizing teams responsible for different components of the assessment may be necessary, depending on the size and scope of the assessment. In the global MA, researchers were divided into three working groups focused on condition and trends of ecosystem services, scenarios, and responses. A fourth working group worked on assembling the sub-global assessments. Sub-global assessments were not usually divided along these lines, and their full assessment teams usually worked together on condition and trends, scenarios, and responses in order to facilitate the integration of these components of the assessment.

Assessment work can be divided by scale (in the case of multiscale assessments), by ecosystem service, or by focal question. In all cases it is important that the members of the assessment team meet frequently in order to ensure that all components of the assessment can eventually be integrated. A lesson learned from the MA sub-global assessments is that it is difficult to integrate findings at the end, and therefore it might be useful to designate a team member to focus on integration issues and to meet regularly to discuss integration of findings. Other roles to be designated within the work team are listed in section 2.3.1. The most effective way to manage the work of the assessment team is described in the next section.

2.4 How to manage the assessment process

Section's take-home messages

- The way participants are convened is a first opportunity to build the legitimacy and credibility of the process, but also a way to undermine it. The convening process is also an early opportunity to start communicating about the assessment.
- At the first meetings, participants need to make some fundamental decisions: What are the precise goals of the assessment? What conceptual approach will be used? What rules will govern decision making?
- A clear workplan will help minimize problems and address issues that may arise during an assessment. Management issues that should be considered include periodic meetings of the technical team and any governing body, stakeholder consultations, conflict management, sources of information, and processes for peer review.
- Capacity-building activities are an integral component of any assessment. The focus of capacity building can differ, depending on identified needs. And the capacities of both those conducting the assessment and those using it might require development.

2.4.1 Convening assessment participants

As noted, assessments may be undertaken upon the initiative of a group of individuals or organizations or even by legal mandate (as State of the Environment Reports are in many countries). However, as mentioned previously, there is always an actor with initiative or leadership who starts identifying the members of the governance structure and the technical teams and brings them together.

Convening is not as simple as it sounds, even if the assessment is undertaken by mandate of a higher authority. The way participants are convened is a first opportunity to build the legitimacy and credibility of the process, but also a way to undermine it. Some important aspects to consider when convening are:

- *Leadership*—Conveners need to be respected by peers.
- *Formality*—Participants must perceive that the process they are being invited to is appropriately managed. The exploratory stage of the assessment may be iterative and flexible, but once the assessment is formally launched, a degree of formality is a strategic way to convince stakeholders of the importance of the work.
- *Representation*—In the case of the governing structure, participants must ideally represent the major stakeholders.
- *Transparency*—The criteria and mechanism for the selection of participants must be clear to all and avoid arbitrary decisions.

A good way for proponents to proceed is to engage first those who will be the leading figures of the governance and technical structures and then to work with them in the gradual identification of the other members. That is, if the governance structure involves an advisory group, a leading figure might be invited to chair it and participate in identifying the other members, preferably with the aid of a completed user needs assessment (see section 2.2.3). As nominations proceed, for reasons of balance or the need to engage a key stakeholder, a co-chair may be appointed.

The convening process is also an early opportunity to start communicating about the assessment. Ideally the process will be perceived as of such high quality and with such potential real impact that stakeholders and experts will find it attractive to participate.

The question of who gets involved in an assessment is a difficult one to deal with, as interest should be the most important factor deciding participation. Genuine interest is what is likely to sustain participation and determine its quality. Representatives of different user groups should be invited to participate, but whether they do decide to participate or not will depend on other issues. Incentives can facilitate the participation process and can help to keep people engaged. The type of incentive needs to be considered carefully in order not to raise expectations unrealistically and to ensure that the right people are attracted. The sustainability of the incentive system also has to be considered.

2.4.2 Early governance decisions: goals, approach, and rules of the process

Once participants have accepted the invitation to participate and their roles have been defined (Chairs of the governance structure, at-large or institutional members of the advisory group, lead scientist on the technical team, etc.), the group needs to make its first fundamental decisions: What are the precise goals of the assessment? What conceptual approach will be used? What rules will govern decision making?

Goals

The assessment must have a clearly stated purpose that will guide the whole effort, from information generation to communication and engagement activities. Because the process involves people with very different backgrounds and understandings, it is important that participants take sufficient time to discuss and define what they are trying to achieve. The goals of various MA sub-global assessments have included:

- Informing a development plan for one region in a country;
- Building a rationale for local management of ecosystem services and landscapes;
- Improving national management of ecosystem services;
- Building local appreciation for ecosystem services;
- Informing a fisheries management plan;
- Understanding trade-offs between agricultural development and other ecosystem services; and
- Developing baseline data on ecosystem services and their relation to human well-being.

Any focused goal is acceptable, but ideally a mechanism should be in place for actors and decision makers to use assessment information to improve the management of ecosystem services. The identification of policies, initiatives, and projects that the assessment can feed into can help in this regard (see section 2.2.3).

Assessment approach and the conceptual framework

Equally important as the goal is a common understanding of the conceptual and methodological approach of the assessment. Not only does this determine the assessment work, it also facilitates the task of communicating the goal and rationale of the initiative to its diverse users. During the exploratory stage and when convening participants, promoters will have presented and discussed the rationale for an assessment exercise. They would have offered the framework as a robust and desirable way to approach the relationship between ecosystems and human well-being. Clearly, however, there might be other views—for instance, indigenous people's approaches to nature and ecosystem services—that might call for changes, either subtle or substantial, to the approach. So the whole approach needs to be discussed and validated by the governance and technical structures. Chapter 3 explains the benefits and approaches to participatory processes aimed at building a common understanding of a particular system through the use of a conceptual framework.

Rules of the process

To ensure credibility and legitimacy, it is important to have clear rules about who will decide things like how information will be generated (e.g., what sources are acceptable or how reviews will proceed), how the reports will be structured, what language to use, what the appropriate communication strategy is, how to set priorities on the use of resources, and when the reports will be officially considered final. It is up to members of the governance structure to decide this as early on in the process as possible and to develop documents that outline these rules. Formalizing the rules will help the assessment team stay on track.

Establishing process rules builds legitimacy because they promote transparency and mechanisms of accountability in the process and because participants will then also be responsible for the correct implementation of the assessment. Credibility, in turn, results from the determination of adequate standards for quality of the scientific or technical work. (See Box 2.6.)

2.4.3 Governance and management issues during the assessment

Being a social process, an infinite number of issues and problems will emerge in the course of any assessment. Some elements can be built into the design of the assessment, however, that will help minimize problems and address the main issues that may arise.

A clear workplan

To be able to monitor progress, it is useful to have a workplan with clearly defined timelines and milestones. (Figure 2.1 presents the MA workplan and timeline.) This is especially important in assessments with several components, where it is necessary to integrate the work of different teams into a single product. Integration is better undertaken as the assessment progresses instead of waiting until the end. The experience of the Southern African Millennium Ecosystem Assessment was that integration needs to be planned from the outset of the assessment, as it is difficult to achieve afterward. In the case of SAfMA, regular meetings of the assessment teams and the Advisory Committee were held to review progress against agreed timelines and to address problems and keep the assessment on course.

Box 2.6. Data quality assurance in the MA

Quality assurance of data is needed in any assessment, and there are different ways of achieving this. The MA adopted the following rules to assure data quality:

- Most data used or cited must be from peer-reviewed scientific publications.
- Most data sets used are from large national or international organizations that have internal procedures for maintaining quality control.
- Datasets developed by the MA will aim at a high level of quality control and archive all metadata.
- Archiving of metadata is designed to help assure the quality of information coming from traditional knowledge and undocumented experience.
- Local and traditional knowledge needs to be critically assessed (e.g., cross-checked or triangulated) before being used.

These rules were applied at the global scale in the MA but were more challenging to apply at subglobal scales where fewer published data were available. However, extremely successful uses of local and traditional knowledge produced comprehensive and credible data, collected through a variety of rigorous approaches. In Peru, communities developed data on conditions and trends in soil and water resources by consensus. In communities in South Africa, researchers and community members used Participatory Rapid Appraisal techniques to develop and validate data on several ecosystem services. Chapter 4 presents different approaches for validating information from a variety of sources.

Source: Georgina Cundill, SAfMA.

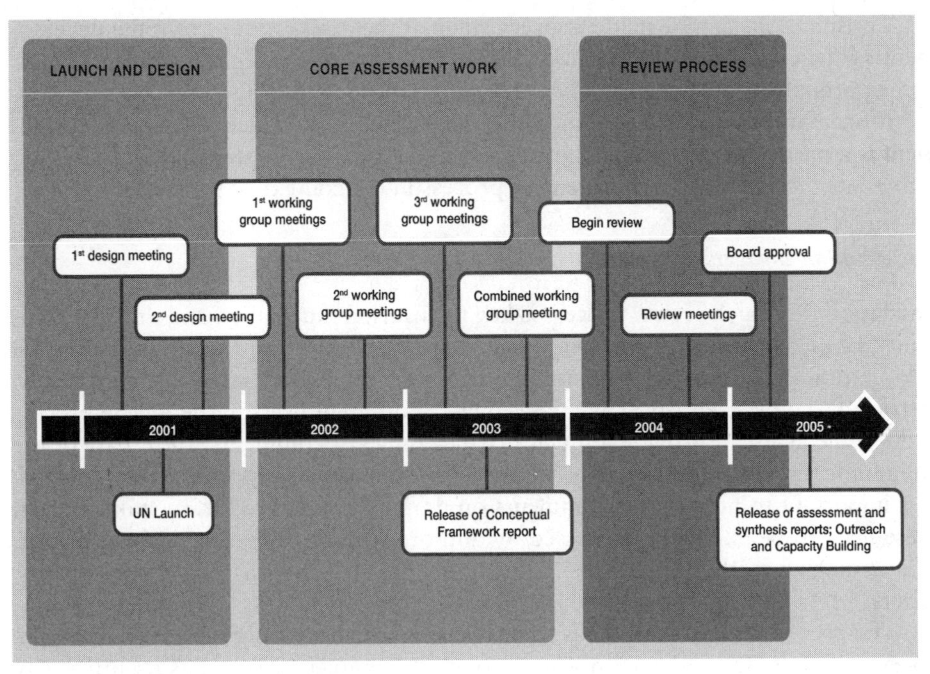

Figure 2.1. Schematic presentation of MA workplan.

Interaction between the technical team and the governance structure

Throughout the process it is critical to have a fluid and positive interaction between those gathering the information and putting together the reports and the group of people guiding the process, who will ultimately "approve" the work. Periodic meetings between the technical team and the advisory group or governing body of the assessment serve various purposes, mostly related to the relevance and legitimacy of the process, such as:

- Checking on the fit between the policy questions posed in the assessment and the technical answers provided;
- Checking that the format of the assessment is useful and friendly for the target audience;
- Checking that the language being used is appropriate for targeted users;
- Identifying the main findings that need to be communicated, and defining how they should be stated; and
- Managing the expectations of stakeholders.

Stakeholder consultations

Consulting with a broader range of stakeholders than those represented in the governing bodies is a good way of validating the latter's decisions, obtaining further input on the technical work, and disseminating findings in advance of the final products. Formal stakeholder consultations occur during the review process, but other workshops and meetings give stakeholders the opportunity to ask questions and voice any concerns they may have.

Consultations can be made at every stage of the assessment—from the development of the conceptual framework to different parts of the technical teams' work. Some parts of the assessment, however, will be more malleable to stakeholder consultation than others, and this should be explained in advance. Scenario development is a particularly useful part of the assessment to engage stakeholders. Chapter 5 presents the stakeholder engagement process in the context of scenarios.

Review as an engagement process

Validation through a review process is key to ensuring the quality, and thus the credibility, of an assessment. Peer review is a standard way of approving the quality of information in the scientific community. However, the review process should not be restricted to only scientists. The involvement of different users in the review process is desirable as it can provide a much broader range of comments, form part of the communication strategy, and contribute to ongoing user engagement in the process. It is important to note that if comments on drafts are requested from stakeholders, these need to be explicitly addressed by authors whether they accept them or not. It is damaging to the process to invite comments and then disregard participants' inputs.

The peer review process needs to be transparent and should be agreed to by the governing body. Review need not focus only on complete reports, as even the way the assessment is progressing and the interim products can be reviewed. In SAfMA, the Advisory Committee reviewed work in progress in addition to completed draft reports.

Conflict management

As in any participatory process, conflicts of interest among stakeholders are very likely to come up, and the advisory group should be well positioned to resolve them. Plenty of tools for conflict management, constructive negotiation, and facilitation can be applied, but it will always be critical that group meetings are adequately organized and chaired. Some ways to minimize conflict during the process are to:

- Establish by consensus clear, but flexible, rules of participation;
- Have an agenda and clear objectives for each meeting that is convened;
- Promote communication among members in between meetings; and
- If the governing body is a large one, create a committee to deal with operative issues between meetings.

2.4.4 Issues in the technical process: sources of information, review, and capacity building

Identifying sources of information

Chapters 4, 5, and 6 point to sources of information for assessing conditions and trends of ecosystem services and human well-being, developing scenarios, and assessing responses. In many situations, particularly in developing countries or in remote locations, information on ecosystem goods and services (beyond provisioning services) is scarce. The exploratory stage of the assessment will usually reveal this condition early on. Depending on the resources, capabilities, and time frame, decisions will have to be made about the need for generating new data. If this is not possible, the advisory group and the project team will have to devise ways for dealing with the lack of data. The systematic use of experts' judgment might be an appropriate way to generate relevant information.

In addition to science, MA sub-global assessments have favored, to the extent possible, the use of local or traditional knowledge and practitioners' knowledge (a comprehensive definition of systems of knowledge is found in MA 2005 and Reid et al. 2006). This, however, might not be immediately accessible and will probably require a dedicated effort to build trust and commitment to the project by the holders of local knowledge. Another challenge will be the eventual integration of local knowledge with published information.

Peer review as a technical process to ensure credibility

As mentioned in previous sections, peer review of assessment findings is a key mechanism for building the relevance, credibility, and legitimacy of the assessment. The credibility of the findings improves with each additional review, especially if the pool of reviewers has a diversity of expert perspectives. The legitimacy of the findings improves as stakeholders are given the opportunity to comment on the findings and take part in the process. The relevance of the assessment is improved as stakeholders can comment on whether their specific information needs are being met. The review process must therefore be conducted in a transparent and professional manner. This can occur anywhere from once to several times during the assessment process. Each additional review helps the process and outcomes, but it requires additional resources and time.

It is common for advisory groups and assessment teams to develop a list of potential reviewers well before the review process occurs. These individuals can be chosen on the basis of their technical expertise, decision-making capacity, or political stake in the region and its ecosystem services. Reviewers can be identified by the experts writing the reports, by the Board, by assessment proponents, or by stakeholders. Reviewers thus often include local people and international experts. Advisory groups associated with MA sub-global assessments were very useful in the review process, as they generally included representatives from many user groups.

Review processes require advance warning for reviewers in order to prepare them for the relatively short windows of time in which they have to read assessment drafts and submit comments. Reviewers should be sent invitations to review either specific assessment chapters (if the assessment is long) or relevant sections of full reports, to prevent overburdening individual reviewers and to maximize the chance of getting a response. These invitations could be sent out several months before the review process begins and be followed up with correspondence when necessary. The dates of the review should be listed from the beginning. When the review period approaches, reviewers are sent a form that they will complete with review comments. (See Figure 2.2 for an example of such a form.)

Review comments can then be sent to the appropriate assessment team members responsible for the specific sections referred to by each comment. Each review comment needs to be acknowledged, either by adjusting the assessment text (if the author agrees with the comment) or by explaining in writing why no change has been made. The transparency of the process is improved if the review comments and author responses are made public.

Capacity building

Capacity-building activities are an integral component of any assessment, but especially a complex one in the style of the MA. In the MA, capacity building served to overcome a variety of constraints faced by a number of sub-global assessments. Many teams did not have the capacity to fully address the linkages between ecosystem services, human well-being, and drivers of change, were challenged in the development of scenarios, and were not able to work at multiple scales. The focus of capacity building can differ, depending on identified needs. The capacities of both those conducting the assessment and those using it might require development.

Different approaches can be used to build the capacity of assessment teams. In the MA, sub-global assessment teams met at MA sub-global working group meetings to share methodologies and lessons learned. Regular workshops can be held at the level of an assessment to build capacity. Invited experts can hold seminars on how to conduct specific technical work. The involvement of students and young researchers at the beginning of their careers is another way of building capacity.

The ability of different institutions and interest groups to use assessment findings may be limited in some cases. In the Bajo Chirripo assessment, for instance, a

[Title of assessment]

Dear Colleague,

Thank you for agreeing to review the [assessment]. In order to ensure that comments from reviews are handled in the most effective manner possible, we would be grateful if you could provide your comments in the table below.

Comments to be returned to [coordinator of assessment] by [deadline]

Name:

Affiliation:

General Comments

Specific Comments

Page No.	Line No.	Review comment	Author response to comment
01	40	Replace "XX" with "YY"	Change accepted
02	01 - 12	The paragraph seems to contradict what is	Paragraph refers to a
		said in page XX...	different issue...

Figure 2.2. Example of a form to request peer reviews.

local NGO attempted to develop resource management plans with local communities based on assessment findings, but the communities did not have established institutions to implement the findings effectively (MA 2005). In some cases, cooperation is required among regional institutions in order to implement the findings of an assessment. Building capacity to use findings can thus range from giving workshops on the conceptual framework or potential response mechanisms to convening different user groups to encourage partnerships that will work together to make use of findings. The capacity of users to apply the findings of an assessment will be enhanced through their sustained involvement in the assessment and their participation in activities to discuss and analyze issues central to it.

2.5 How to bring the assessment to a close

Section's take-home messages

- The assessment can be considered finished once it has undergone a sufficient number of peer and stakeholder reviews and has been approved by the technical team or by the leading authors.
- The final discussion to approve the assessment findings is extremely important for the assessment's legitimacy. This is also a final opportunity for any stakeholders who see their interests threatened by the assessment findings to voice their opposition.

An assessment can be expected to raise more questions than answers. There will always be a need for further research and analysis. However, at some point both the technical team and the stakeholders involved will have to express their satisfaction, or lack thereof, with the "final" work. So when can an assessment be considered finished?

From the perspective of the technical work, the assessment can be considered finished once it has:

- Undergone a sufficient number of reviews (both peer reviews and stakeholder reviews)—Determining how many reviews are sufficient depends on each particular assessment, but normally at least two rounds are advisable.
- Been approved by the technical team as a whole or by the leading authors—Open, technical scrutiny by everyone of everyone else's work will help the quality of the work and allow team leaders to sign off the final document.

Once these two conditions, which basically consolidate the credibility of the work, have been met, the technical team can present the final document for discussion and approval by the governing body of the assessment, where stakeholders will decide whether the work is officially approved or not.

The final discussion to approve the assessment findings is extremely important for the assessment's legitimacy. If interaction between the technical team and the governing structure has been fluid throughout the process, this final discussion should not present major problems. But this is also a final opportunity for any stakeholders who see their interests threatened by the assessment and its findings to voice their opposition to it.

Normally, the documents of most concern for stakeholders will be those that will be widely distributed and used for communication: the summaries for decision makers or the thematic syntheses. There is wide scope for interpretation between the way findings are produced by scientists and a set of conclusions that can be effectively communicated to stakeholders and the wider public. Different scientific findings can be emphasized in different ways, spins can be put on statements, narratives can be constructed by collecting findings from different sections of the assessment, and so on. The discussion is thus a critically important one and must proceed with the active participation of the technical team, who must tread a thin line between preserving the "scientific truth" of the findings and yielding to the political priorities of users.

This has been the case with the Intergovernmental Panel on Climate Change, whose results are debated by an assembly of government representatives. Governments approve the work of IPCC, but they focus long and hard discussions on the summaries for decision makers, the contents of which are scrutinized line by line and often dramatically altered from what the technical team originally proposed.

With the formal approval of the documents by the governing bodies, the assessment findings are now ready to be widely disseminated.

2.6 How to communicate assessment findings

Section's take-home messages

- Effective dissemination of results needs to be guided by clear communication goals. To develop a strategic communication plan, it is important to know the target audience or audiences well. One way to do that is to identify all the people and organizations in a position to influence the types of interventions that have been identified as desirable in the assessment.
- Deciding on the key messages of an assessment is one of the most important steps of the communication process—messages that are a strategic culling of the points most relevant to each audience, presented in a way that promotes the credibility of the findings. Experience has shown that assessments may lose legitimacy and credibility in the eyes of some audiences if they go beyond presenting objective scientific findings into suggesting how policy makers should do their jobs.
- The strategy and extensiveness of a communication program will depend on the assessment's budget. The formats for presenting findings include early release of products such as the conceptual framework, final reports tailored for specific audiences, summaries for decision makers, Web sites and other online resources, workshops and meetings, and coverage in various media.

2.6.1 Defining a communication goal and knowing the audience

An effective dissemination of results needs to be guided by clear communication goals. This is the goal that will define the specific target audiences, which will determine the appropriate means of dissemination.

The dissemination or communication goal should be subservient to the general goal of the assessment, and dissemination activities should all support the purpose of the whole assessment. It is important to keep this in mind because it helps discriminate between the broad range of good communication ideas to focus resources on those that are most specifically conducive to the substantive goal.

In the case of the MA, the purpose of the assessment was to establish the scientific basis for actions needed for the conservation and sustainable use of ecosystems for human well-being. To help achieve this, the dissemination strategy focused on two goals: ensuring that stakeholders were adequately engaged through appropriate access to the assessment process and its products and creating a demand for the assessment reports and for technical expertise to conduct sub-global assessments. It was understood that these two dissemination goals were key for achieving the main goal of the assessment.

In order to develop a strategic communication plan, it is necessary to know the target audience well. The audience will include the stakeholders that have been involved in shaping the assessment from the beginning (see section 2.2.3), but potentially many other groups, such as international organizations, businesses, and public officials in a diversity of sectors, as well as the larger public. In the MA sub-global assessments, the diversity of audiences was often defined by the scale of the assessment, with larger regional assessments having a higher diversity of audiences and a broader communication strategy. One way to determine the target audiences is to identify all the people and organizations in a position to influence the types of interventions that have been identified as desirable in the assessment and then to reduce this list based on what is a feasible strategy under time and budget constraints.

Common audiences for assessment information include:

- Governments (various levels and various departments)
- Planners
- Politicians
- Researchers and analysts
- Nongovernmental organizations
- General public
- Schools and universities
- Industries and businesses
- Women's groups
- Indigenous peoples' groups
- Media.

Target audiences are defined by their profession and areas of focus, which will influence the content and style of the materials used to reach them. Audiences are also defined by differences in language and culture. This may result in increased costs of printing materials in multiple languages or the need to have several strategies for the dissemination of assessment findings using very different media. In some areas, for example, newspaper articles may be effective for disseminating information to the general public, while in other areas the radio or television may work better. In still other cases, personalized approaches and targeted products are required.

While putting together a list of desired audiences for the assessment information, it helps to take notes on what kind of information might be most relevant or useful to different audiences, the perceptions of each group on the issues included in the assessment, and the type of communication method has been used to reach each group in the past. It is also useful to distinguish between potential end users of assessment information, who make decisions based on the information (e.g., adopt a law or not, buy or not buy), and "broadcasters," who recycle information for their own communication goals and thus multiply its impact (e.g., the mass media, the educational system, and many NGOs).

2.6.2 Developing the content and style of reports

Deciding on the key messages of an assessment is one of the most important steps of the communication process. Full assessment reports are useful reference documents and will contain all the information produced during the assessment. But these documents will rarely be used to disseminate information to target audiences. At this point, the content and conclusions must be synthesized into short and specific messages that will resonate with the audience. The main messages are usually not simply a summary of all the information produced by an assessment but rather a more strategic culling of the points most relevant to each audience, presented in a way that promotes the credibility of the findings. This means backing up important statements with data and examples and using easy-to-understand graphs, illustrations, and tables. Graphic figures can be very powerful tools for conveying complex information in a way that is understandable and memorable.

The MA took care that outputs developed for communication were, like the assessment itself, relevant to policy makers but did not tell them what to do. In other words, main messages were policy relevant but not policy prescriptive. This is generally because assessments may lose legitimacy and credibility in the eyes of some audiences if they go beyond presenting objective scientific findings into suggesting how policy makers should do their jobs. Assessments can still affect policy by making sure that the information that is most relevant to the choices being faced by audiences is included in the communication products.

The style in which the main findings are communicated depends on the audience. Box 2.7 gives some examples of how to match the style of reporting to a specific audience. Section 2.6.3 outlines the different formats that can be used to communicate findings. But within each format, attention to the style of presenting information is key to reaching the target audience.

Acknowledging uncertainty is also a strategic part of putting together messages that preserve the credibility of the work. The assessment team must decide whether to include information that is uncertain. If it is included, the associated level of uncertainty needs to be clearly stated. As described in Chapter 1, the MA assigned certainty levels to findings based on the collective judgment of the authors, who used observational evidence, modeling results, and theory to decide which level of certainty applied. (See Box 1.3 in Chapter 1 for further details.)

2.6.3 Communication formats

The strategy and extensiveness of a communication program will depend on the assessment's budget. With a clear idea of the communication goal, the target audiences, their information needs, and the available budget and communication expertise, the next step is to decide on the format of communication. This section describes several common formats, but many others may be suitable.

Box 2.7. Target groups and report style

- *Decision makers*. Content should be short, specific, fact based, and consist of the latest information.
- *Media*. Content should be short and consist of findings relevant for broad audiences, with messages that can easily be linked to other issues in the news. There is a better chance of media coverage if there are supporting visuals such as graphs or photographs.
- *Students*. Content should be well explained, and the language should be simple.
- *Scientists*. Content should be fact based and rely on the latest data. The language can be scientific and include technical terms.

Source: UNEP 2007.

Tailored reports

To ensure that as many people as possible could obtain access to the information, the MA produced reports tailored to different audiences. This approach involves writing targeted publications that focus on the most relevant information for specific groups and sometimes translating publications into multiple languages. In addition to the full technical assessment reports, the MA produced six synthesis reports aimed at different users. Information from the main assessment volumes was summarized and repackaged in short, carefully designed volumes dealing specifically with biodiversity, desertification, wetlands, health, and business and industry, in addition to an overall synthesis directed at a more general audience. Content, language, and style were modified to suit each audience group.

When designing a report for a particular audience, it helps to ask the following questions:

- *How long a document will a person in this position typically read?* Often a decision maker will not read more than 1–5 pages, while their advisors might read 10–30 pages. Only scientists and practitioners will read longer, technical reports.
- *Should the content be written in technical, popular, or formal policy language?* The document must be both appealing and easy to understand for the intended audience.
- *What kinds of issues or decisions are facing this audience?* Is there a specific issue or decision that the report can address explicitly? The report will ideally not offer guidance on a decision but rather will supply targeted, relevant information.
- *What information from the assessment needs to be included?* In some cases, only a small segment of the entire report will need to be presented to a particular audience.
- *What figures and formatting will be effective in communicating to this audience?* The presentation of the information, including clear figures, graphs, and drawings, and the layout of the document will make a big difference in how different audiences take up the information. It is often helpful to look over other reports and discuss with users what they find effective and appealing.

Depending on the scale and scope of the assessment, the reports may need to be translated into several languages. Although in the end almost all the synthesis reports were translated into the five U.N. languages, this proved to be one of the main dissemination difficulties in the MA, with translations taking longer than desired and requiring multiple reviews to ensure quality. Leaving certain language groups out by not translating reports can politically undermine the dissemination process. Translation of reports is an opportunity to engage more stakeholders in the process, such as universities and NGOs, which can then also assist very effectively in dissemination activities. Hence it would be useful to begin engagement arrangements for translation early on in the process.

Summaries for decision makers

A category of tailored report is the summary for decision makers (SDM or SPM [summary for policy makers]). This is usually a very short document (one to several pages) that highlights the key messages in one or two sentences each. It does not provide all the evidence behind the findings, but there should be a clear indication of where further information can be found. When messages are boiled down to this degree, there is a tendency to make very general statements that may not relate directly to the policy action agenda of the targeted decision maker. Going over the main messages with decision makers is one way to come up with precise, strong messages that can be acted upon. But even when only general statements can be made, assessments can be very significant sources that help policy makers reinforce an argument, confirm a widely held belief, or contest it. When the MA, in its summaries for decision makers, said that "over the past 50 years, humans have changed ecosystems more rapidly and extensively than in any comparable period of time in human history, largely to meet rapidly growing demands for food, fresh water, timber, fiber, and fuel," it was not revealing something that many people did not intuitively know. But the weight of this statement expressed by hundreds of the best scientists in the world turned it into a powerful communication instrument.

Electronic communications

Establishing electronic communication mechanisms is important, as a growing number of people find and share information on the Internet. Electronic communication mechanisms may include a Web site, a system to share data, and an intranet system for internal communications among the assessment team members. Electronic mechanisms are becoming easier to set up, even without programming expertise, although a well-designed Web site and intranet often require the input of a specialist. Assessments with larger budgets may choose to hire a Web designer to produce a professional, easy-to-navigate site where all the assessment products will eventually be located. Smaller projects can take advantage of free online resources, including preformatted Web sites, blog spaces, and wikis, among others. Local universities are a good resource for locating ideas, expertise, and available Internet sites.

Workshops and meetings

Explaining the findings of an assessment in person is a powerful way of disseminating information. Workshops and meetings provide audiences with the opportunity to ask questions and understand the findings more deeply than they would through reading a report. However, the trade-off is that these meetings are costly and only reach a small number of people. At smaller scales, this may be the most effective communication strategy and does not have to be too costly. For larger-scale assessments, the most relevant decision makers and stakeholders can be invited to a workshop to discuss assessment findings, while other methods may be used to disseminate information to the broader public. Workshops can also be used to build capacity among different stakeholders to communicate the principal messages to other audiences.

Nontraditional communication methods

There may be an opportunity to use less traditional means of communication—theater, art, calendars, or video, for example—to capture an audience and communicate important messages to them in imaginative ways. Some care should be taken to ensure that the credibility of the findings is not compromised by the method of communication, as some audiences will be more receptive to nontraditional methods of communication than others. In some cases, these methods will be more appropriate or effective than reports. In addition to reports and summaries, the MA sub-global assessments produced brochures and pamphlets, atlases, posters, calendars, theater pieces, and videos. For example, the local assessment in Vilcanota, Peru, trained community members to produce a video to disseminate findings to local communities. Video was considered to be culturally appropriate in this context, as it is a form of communication that is similar to the local tradition of storytelling and visual representation of environment and culture.

Media

The media plays an important role in disseminating assessment findings to the general public. Assessment teams can prepare press releases and make the media aware of people who are available to answer questions. The MA was a high-profile international assessment and wanted to achieve a certain level of press attention when the findings were finally released. Three approaches were used to accomplish this:

- Organizing seminars for the media while the assessment was being conducted to explain what it was, why it was being done, and what to expect from it;
- Establishing a loose working group with the media officers of partner organizations; and
- On the day the MA was released, organizing press briefings and seminars in 13 cities around the world, which ensured that appropriate angles and languages were used to draw national media attention.

The MA also posted a list of contributing authors around the world who were available for media interviews. This allowed local press to contact authors in their areas who could link the MA findings to local issues.

2.6.4 Conveying relevance, credibility, and legitimacy to the audience

The relevance of the final assessment products will depend on how well the stakeholder process was set up from the beginning. However, having different communication products tailored to different audiences will help maintain the relevance of the assessment to those audiences, and they will appreciate being able to read only the information they would be most interested in.

There are many approaches for branding assessment products as credible and legitimate:

- Have enough participation and buy in from well-known organizations to put their logo on the assessment products;
- Have highly respected scientists, politicians, or public figures introduce the assessment to the media and general public;
- Invite respected and well-connected people to join the advisory group and help communicate the assessment findings; and

- Organize a thorough review process and include the number of reviewers and review comments on communication products.

2.6.5 Strategies to leverage communication

Champions

In many of the MA sub-global assessments (and also at the global level), specific individuals played key roles during different stages of the assessment—for example, as external facilitators in determining the demand for the assessment, in providing leadership and sustaining the process, and in communicating findings. In some cases this was a member of the assessment technical team; in other cases it was someone who played an advisory role. Many sub-global assessments found advisory group members to be a powerful means for communicating the findings to a diversity of audiences. Some advisory group members might therefore be chosen specifically for this purpose and be high-profile, respected individuals within a particular context.

Champions can also be individuals who are highly respected and considered to be neutral within politically charged or conflicted contexts. In these cases, the individual is usually not associated with the assessment but can help build trust among audiences in order to communicate the findings effectively.

Partner institutions

The engagement and outreach team of the MA saw its communication activities as an instrument not just to reach out and convey an image of the MA but also to enhance the ownership of the MA and improve the ability of third parties to understand and make better use of it. Hence, the MA sought to rely on as many partners as possible for outreach and to encourage many third parties to undertake outreach for the MA on their own. This resulted in several instances where volunteers approached the MA to undertake activities, which was highly beneficial in dealing with media enquiries.

The MA found that partnering required two important elements: a minimum level of coordination in terms of setting key dates and sharing basic strategies for communicating assessment findings, plus the generation of materials to support outreach by third parties. These materials were shared not only through the MA intranet but also through an "outreach kit" distributed on CD. This contained a collection of elements developed by the MA, including:

- Guidance on how to explain the MA to the uninitiated;
- Guidance on how to develop a communications strategy;
- Graphic elements (posters, maps, logos, photographs, videos); and
- PowerPoint slides.

While all assessments might not be able to produce as complete a kit, the same types of products and guidance can be shared through a simple document or developed in a workshop with the partner institutions.

Sustained interaction with audiences

Keeping an assessment visible for its potential users is key to building up enthusiasm for the findings. Often assessments are launched with the participation of stakeholders and then the whole process takes up to several years to complete. For example, the global MA was a four-year endeavor. Even after being approached and consulted at the inception of the assessment, targeted users needed to be kept updated and reminded of upcoming work. These activities were also meant to build momentum and expectations; they involved multiple briefings and smaller meetings in international and national arenas. Briefing audiences on the progress of the assessment is a simple way to maintain a positive relationship with audiences.

Early products

One way to increase the demand for assessment findings is to release some products early. The global MA did not wait until the end to start releasing outputs, for example. In particular, releasing the conceptual framework and early findings on sub-global assessments permitted better outreach during the process. The Portugal sub-global assessment released a User Needs report near the beginning of its process that showed how the assessment would meet the information needs of important national stakeholders. Early products can pique the interest of audiences and give them a concrete example of what can be expected at the end of the assessment.

The assessment provides diverse opportunities for generating and disseminating relevant information. Most of the assessment elements, such as conditions and trends on ecosystems and human well-being, scenarios, and responses, might be valuable stand-alone products for users. A timely dissemination—for instance, first as drafts for comments and then as final products—will secure the interest of users and increase their constructive involvement. If the assessment team includes the dissemination of intermediary products in their workplan (e.g., a status report), complying with this plan will be seen as a positive signal of the project's success.

References

Agrawala, S. 1998. *Structural and process history of the intergovernmental panel on climate change.* Climatic Change 39(4): 621–642.

Cash, D., Clark, W., Alcock, F., Dickson, N., Eckley, N., Jager, J. 2002. *Salience, credibility, legitimacy and boundaries: linking research, assessment and decision making.* John F. Kennedy School of Government, Harvard University. Faculty Research Working Papers Series. RWP02-046.

Eckley, Noelle. 2001. *Designing effective assessments: the role of participation, science and governance, and focus.* Report of a workshop co-organized by the Global Environmental Assessment Project and the European Environment Agency, 1–3 March, Copenhagen, Denmark. Research and Assessment Systems for Sustainability Program Discussion Paper 2001-16. Cambridge, MA: Environment and Natural Resources Program, Belfer Center for Science and International Affairs, Kennedy School of Government, Harvard University.

Farrell, Alexander E., and Jill Jäger, eds. 2005. *Assessments of regional and global environmental risks: designing processes for the effective use of science in decisionmaking.* Washington, D.C.: RFF Press.

MA (Millennium Ecosystem Assessment), 2003. *Ecosystems and Human Well-Being: A Framework for Assessment.* Island Press, Washington, D.C., 245 pp.

MA (Millennium Ecosystem Assessment). 2005. Ecosystems and Human Well-being, Volume 4: Multiscale Assessments: Findings of the Sub-global Assessments Working Group. Island Press, Washington, D.C.

Mitchell, Ronald B., William C. Clark, David W. Cash, and Nancy M. Dickson, eds. 2006. *Global environmental assessments: information and influence.* Cambridge: MIT Press.

Pahl-Wostl, C., 2003: Polycentric integrated assessment. In: *Scaling Issues in Integrated Assessment,* J. Rotmans and D.S. Rothman (eds.), Swets & Zeitlinger, Lisse, the Netherlands, pp. 237–261.

Ranganathan, J., Raudsepp-Hearne, C., Lucas, N., Irwin, F., Zurek, M., Bennett, K., Ash, N., West, P. 2008. *Ecosystem Services: A Guide for Decision Makers.* World Resources Institute.

Raustiala, K. and D. Victor. 1996. Biodiversity since Rio: The future of the Convention on Biological Diversity. *Environment* 38(4): 17–20, 37–45.

Reid, W. V., Berkes, F., Wilbanks, T. J. & Capistrano, D. (2006) *Bridging scales and knowledge systems: concepts and applications in ecosystem assessment* (Island Press, Washington D.C.).

3

Conceptual Frameworks for Ecosystem Assessment: Their Development, Ownership, and Use

Thomas P. Tomich, Alejandro Argumedo, Ivar Baste, Esther Camac, Colin Filer, Keisha Garcia, Kelly Garbach, Helmut Geist, Anne-Marie Izac, Louis Lebel, Marcus Lee, Maiko Nishi, Lennart Olsson, Ciara Raudsepp-Hearne, Maurice Rawlins, Robert Scholes, and Meine van Noordwijk

---

What is this chapter about?

This chapter provides information on and lessons from experiences with conceptual frameworks that may help in adapting and developing a framework for an ecosystem assessment. The social process to create the conceptual framework is as important as the final product. This creative process requires interaction—and often involves tension—between users and the assessment team. The challenge of working together to create a shared conceptual framework can play an important role in creating ownership by the users of the assessment and in building an assessment team.

Recent experiences with global assessments, such as the Millennium Ecosystem Assessment (MA), show that conceptual frameworks can provide greater focus on key issues and relationships and serve a useful role in synthesis and cross-site comparisons. Although the MA framework has in some respects become a standard point of departure for ecosystem assessment, there is no unified theory on creating conceptual frameworks. Examples from MA sub-global assessments illustrate a range of pragmatic approaches, ranging from adaptation of the global conceptual framework to independence from it and including the use of multiple frameworks.

The people who are (or are not) informed about, consulted, and involved in creation of the conceptual framework and the ways in which their knowledge and expertise are valued (or not) will in many ways govern the entire assessment process. Both the groups consulted and the components that are valued by the assessment team as well as the quality of interaction between the assessment team and the stakeholders are important to developing a conceptual framework that effectively balances the principles of legitimacy, relevance, and credibility discussed in Chapter 2.

The chapter begins with a simple definition of a conceptual framework and then discusses some practical considerations of its meaning in ecosystem assessment. Section 3.2 explores the often intertwined challenges and opportunities involved in developing a conceptual framework. Sections 3.3 and 3.4 juxtapose the dual roles of conceptual frameworks in ecosystem assessments: as a means for clarity, credibility, and comparison and as a tool for engagement, usefulness, and legitimacy. Rather than adopting a conceptual framework entirely "off the shelf," a pragmatic approach that blends various frameworks and methods to balance strengths and offset weaknesses seems to be the most appropriate method.

3.1 How to understand the relationship between people and nature

Section's take-home messages

- In ecosystem assessment, a conceptual framework is a concise summary in words or pictures of relationships between people and nature—in other words, among the key components of interactions between humans and ecological systems.
- Conceptual frameworks can help clarify and focus thinking about complex relationships, including how those relationships may be changing over time. They also can be a focus for interaction to build shared understanding.
- The understanding developed when building a conceptual framework includes but is not limited to scientific knowledge. Personal experiences, history, cultural practices and values, political savvy, and other forms of knowledge are also important in clarifying and enriching shared understanding.
- It is possible, perhaps even desirable, for an assessment team to use more than one conceptual framework. What is important, though, is that at least one conceptual framework must be embraced (that is, "owned") by the assessment team and users alike.

This chapter proposes a simple working definition of a conceptual framework for ecosystem assessment: a concise summary in words or pictures of the relationships between people and nature, including how those relationships are changing over time. These frameworks often are anthropocentric, centering on people and their needs. This is because assessment users often are focused on the issue of how human well-being is influenced by environmental change, but also because any effort to change the system will necessarily involve actions to change human behavior. But there are many ways to view relationships between people and nature, and cultural perspectives differ significantly on the centrality of nature or people. These contrasting perspectives can be enlightening in and of themselves; one reason for taking a pluralistic approach to conceptual frameworks is that the choice of frameworks and comparisons among them may help clarify underlying assumptions and the implicit values being placed on different outcomes for people and nature.

Conceptual frameworks can help organize thinking and structure the work that needs to be accomplished when assessing complex ecosystems, social arrangements, and human–environment interactions. From a scientific perspective, a conceptual framework can be viewed as a model to guide the assessment process. Like any model, the framework will always be a simplification; often an extreme simplification. But simplification is a useful tool—indeed, an indispensable tool—in clarifying and focusing an assessment process. This capacity for illuminating abstraction is a key reason why science is indispensable to a credible ecosystem assessment. As discussed in greater detail in section 3.3, a conceptual framework can be helpful to focus on key issues among the myriad natural and social processes affecting ecosystems and human well-being, to frame those issues spanning multiple scales across space and time, and to manage the interlinkages among these elements.

Highlighting underlying assumptions and gaps in understanding is an important part of a well-designed conceptual framework. However, a framework is more than just a list of shared assumptions. Since the assessment framework necessarily is developed by diverse stakeholders, it should ultimately be a synthesis of various ways of knowing, explaining, and valuing ecosystems and human–environment interactions. Developing a shared conceptual framework may generate dialogue among groups with different assumptions, ways of understanding, and approaches to managing dynamic natural and social systems. Therefore, a conceptual framework draws on a variety of types of knowledge. Indeed, it typically will be a synthesis of more than one way of knowing or understanding and hence may not be perfectly consistent or particularly elegant.

The focus necessary for a successful assessment should reflect what people value most about an ecosystem. Different stakeholders may emphasize different parts of the system or different relationships. Engaging diverse groups and assessment users in development of the conceptual framework is important for ensuring that the result is accepted or "owned" by users and by the assessment team. A conceptual framework that is not developed through engagement with participants and stakeholders but is instead "imposed" on those who have to use it can turn out to be a rather fruitless exercise (see Box 3.1).

Do clarity and focus require total commitment to a single framework? No, not at all. Ecosystem assessments have been undertaken successfully with more than one conceptual framework. In the Millennium Ecosystem Assessment, for local assessments in the Southern Africa region two additional conceptual frameworks—on adaptive renewal (Gunderson and Holling 2002) and sustainable livelihoods

Box 3.1. The need for "ownership" by assessment teams

Assessment teams working on the International Assessment of Agricultural Science and Technology for Development (IAASTD) initially intended to use a conceptual framework inspired by the Intergovernmental Panel on Climate Change (IPCC) and the Millennium Ecosystem Assessment (MA). Many authors pointed out what was missing from this framework and essentially used their own approaches in early drafts. As a result, the fact that authors were expected to produce an assessment of options became blurred, as people could not see how to conduct an assessment of agricultural knowledge, science, and technology within the proposed framework.

Eventually the scenarios chapter in the original structure was dropped altogether, which marked the de facto abandonment of the initial conceptual framework. It was not until the different lead authors for different chapters started discussing these challenges directly that a new framework based on the concept of "multifunctionality of agriculture" was developed out of an initiative that originated with the French government. This concept took on a new life over the three years it took to produce the IAASTD, signifying that agriculture provides multiple services to people—from food, feed, and fiber to aesthetic landscapes and ecosystem services. This framework was used (at least in passing) in most of the final chapters.

(Carney 1998)—were "superimposed" and used in a complementary manner with the MA conceptual framework, in order to better capture the dynamic interplay between ecosystems and humans at the local level (MA 2005b:73). The crucial caveat suggested by experience is that whether one, two, or several frameworks are used, there must be at least one that is embraced by the assessment team and users alike.

3.2 How to develop a conceptual framework

Section's take-home messages

- Rather than adopting a conceptual framework entirely "off the shelf," a pragmatic approach for each assessment can blend various frameworks and methods to balance strengths and offset weaknesses.
- It is important to be aware of differing perspectives and conflicting interests within the assessment team and among stakeholders and intended users of the assessment.
- The process of developing a conceptual framework involves predictable lines of tension—among stakeholders, for example, or between the local assessment team and stakeholders, within the local team, or among the local team and assessment colleagues working at coarser scales. It is important to recognize these tensions, as they are not dangerous if dealt with in a respectful setting.
- It is impossible to capture all of reality. A well-constructed conceptual framework should clearly characterize the attributes of the system (conditions and interactions) that are perceived to be most important from the standpoint of the users of the assessment as well as the assessment team.
- To span boundaries among groups, it is important to make efforts to translate, communicate, and mediate across participants who hold different views and have conflicting interests. Semantics matter: the meanings of key words and concepts need to be understood and broadly acceptable.
- The right professionals, with the right skills and experience, need to be involved in the process. In addition to ecologists, economists, and physical scientists, skilled social and political analysts with local knowledge and experience are essential for grounding the process in local reality. Professional facilitators also can help with process insights and techniques for communication and mediation to bridge lines of tension; professional facilitation of the process of creating a conceptual framework can be an excellent investment of time and funds.

The primary concern in this section is a local or national group that is considering undertaking an ecosystem assessment. If they decide to proceed, what positive or negative insights could the team draw from the experience of the MA and other assessment processes? As emphasized in Chapter 2, meeting users' needs is an essential element of a useful assessment process. Similarly, relevance of conceptual frameworks to those users' needs must be addressed as a prerequisite to articulation of broader comparative, synthetic, or "global" aspirations of assessment processes. This chapter—indeed, the entire manual—is intended to help new assessment teams avoid pitfalls and enhance their effectiveness in bringing in local perspectives on "truth" to enrich or at least balance those of "global experts" and "international processes." This section reviews a few lessons and insights regarding challenges of interaction among diverse people with differing interests in order to create an (inevitably) abstract conceptual framework.

3.2.1 Striving for a workable framework that can be understood and is flexible

One of the most important functions of a conceptual framework is as a guide to what is not being done. Since perfection is impossible, pragmatism should be the guide. A workable, operational framework with essential elements is good enough. Remember the saying that "if *everything* is important, *nothing* is important." Although it is important to find the few key links among the possible relationships, it is not helpful if the arrows indicating links show that everything is connected to everything else.

Expect an interesting conceptual framework—one that takes some risks—but also expect it to evolve during an assessment process. Documenting the modifications that arise during this evolution can help new entrants to an assessment process understand better why the current perspective has been adopted. It also can allow those involved to better appreciate what has been learned. A good illustration of the evolution of a particular family of conceptual frameworks is given in Waltner-Toews and Kay (2005) based on their work assessing ecosystem sustainability and health.

3.2.2 Getting the process started

To get started, consider a very simple case as a point of departure: a single person (hence a unified point of view and a single set of values, interests, and objectives) in an isolated, island setting. The steps in developing a conceptual framework for assessment in this kind of "Robinson Crusoe" assessment would include the following:

1. Identify the key elements of well-being or "quality of life," whether or not they are shaped by ecosystem services.
2. Identify the ecosystem goods and services that matter the most for the elements of well-being in (1).
3. Sketch a diagram of the factors that directly affect the supply of the ecosystem goods and services in (2): these can be called "direct drivers."
4. Then, move back one level and add the "indirect drivers" that influence the direct drivers included in (3).
5. Finally, see if there are any connections from the elements of Robinson's well-being in (1) back to the direct drivers (3) or indirect drivers (4) you identified.

The diagram created usually takes the form of "boxes and arrows" (see section 3.3), where the boxes are filled with "things" such as an ecosystem service or a process (population growth), and the arrows are influences from one thing to another. Of course, by framing this as a "one man show," this simple case misses a central point. It is not the ecosystem services that you (as an assessment team member) consider important that matter most. Instead, in a real application, the conceptual framework should feature the services that your intended audience of assessment users considers most important.

The initial diagram often is quite complex. It helps to simplify it to just the most important links—those factors that are changing at a timescale relevant to the assessment. Cut out the factors that are constant or change very slowly and those that fluctuate much faster than the assessment time period. It is also safe to ignore processes that are relatively weak.

The next step is to consider the spatial scale at which key influences occur: Are they imposed at scales much larger than the assessment? Or do they bubble up from much finer scales (smaller than the minimum resolution of the assessment)? The large-scale influences are important, and they can be clarified by handling them as indirect drivers (step 4 above). The finer-scale processes may involve too much detail to be appropriate; they might be eliminated to help simplify the conceptual framework (and the entire assessment task).

The key distinction between this fictional island of Robinson Crusoe's and a more realistic assessment situation is not its ecosystem complexity (which in principle could match that of any real ecosystem) but the social, cultural, and political complexity arising from the multiplicity of different points of view, types of knowledge, values, interests, and objectives of individuals and groups in any society (Reid et al. 2006). The conceptual framework developed in this way is a helpful place to start the larger process, with the understanding that it will change. Building on the engagement strategy described in Chapter 2 to ascertain users' needs, the communications involved in developing a conceptual framework should involve diverse stakeholders and encompass a variety of perspectives and types of knowledge. These connections can help build the shared understanding needed for a successful collaborative process.

3.2.3 Anticipating predictable lines of tension

If a conceptual framework is developed properly, the ecosystem services that become the focus of the framework involve high stakes and, as a result, tend to be politicized, "hot button" issues. These often are the focus of conflict between different stakeholder groups over control and access to valuable resources and over who bears costs and reaps benefits. But the salience of those issues for assessment users may not fit well with the methods and approach that the assessment team needs to ensure credibility. Moreover, because assessment teams must be multidisciplinary in a broad sense (involving social scientists and natural scientists), an additional line of tension typically arises within a team from interactions between scientists (regardless of their discipline) who may have little experience with (or even respect for) colleagues trained in other disciplines. Finally, if an assessment team is engaged with colleagues conducting parallel assessments at coarser scales, the experiences with the global and "sub-global" components of the MA suggest that there could be tension between the team's broad imperative to "get on with it" and the organizer's responsibility to deal with complex, even messy, local reality in order to produce useful results for people with a direct stake in the assessment topics. Tomich et al. (2007) document insights from one process of building a multiscale, multidisciplinary, multicultural team.

3.2.4 Impossibility of a "neutral" or value-free conceptual framework

By definition, a conceptual framework seeks to draw attention to a subset of components and relationships that are believed to be most important for understanding the system. What is "most important" to one person may be less important to others, however. From the standpoint of an environmental activist or government negotiator, the "most important issue" may be a global consideration like greenhouse gas emissions, while from the standpoint of a farmer, it may be access to local resources, such as land or water. As a result, any process to develop a conceptual framework is inherently political, involving balance and contention among different interests and concerns.

Thus all conceptual frameworks are part of the contest of values and interests of different people; some are powerful, others not. Because no one is immune to these pitfalls, it is important for the assessment process to be open to critical examination of biases and prejudices cloaked as "science" or "best practice." Work on agroforestry in Kenya by Jerneck and Olsson (2007) provides an example of how views held by scientists can be a barrier to understanding opportunities not just for local, private benefits but also for enhancing the public good globally. Thus, openness to alternative perspectives applies with particular force to the words and slogans used by the assessment team, who must be sensitive to differing and conflicting views when developing a framework for assessment (see Box 3.2).

This essential sensitivity to alternative views is especially important in the early phases of the assessment process, including development of the conceptual framework. Different people can have very different conceptual starting points (i.e., their own mental models about how the world works). The risk is that the assessment team's uncritical use of words and slogans—and belief in them—can inhibit the process of identifying other views that may be equally valid, relevant, and useful. If handled effectively, then, the diversity of views is not just of academic interest but can be an important asset in creating a framework for assessment.

3.2.5 The process is as important as the framework

The conceptual framework for an assessment depends on purpose, time, and place. It can play a key role in communicating the results of the assessment, especially to audiences that are more visually oriented than word or text focused. But the framework should be seen as the product of a process that has its own value rather than simply as a starting point for assessment. If the framework is derived from a process that truly involves multiple stakeholders and that respects and embraces differences in their perspectives, experience, and knowledge, it can provide an important element of the legitimacy and usefulness required for a successful assessment.

Given this central importance of the conceptual framework both as a process and as a product, and given also the fact that a high-quality process must grapple with social and political realities, it is important to involve the right professionals, who have the right skills and experience. In particular, skilled social and political analysts with local knowledge and experience are essential. And professional facilitators also can help with process insights and techniques for communication and mediation to bridge lines of tension. Indeed, if funding allows, professional facilitation of the process of creating a conceptual framework can be an excellent investment, producing rewards throughout the assessment process.

Box 3.2. Using conceptual frameworks for conceptual clarity

Multiple perspectives and conflicting political interests affect use and interpretation of words within science, just as they do in the "real" world. A single word or slogan sometimes embodies a way of thinking about relationships between people and nature. Whether used consciously or (more often) unconsciously, these concepts can be packed with complex meanings for one person that are neither shared nor apparent to others.

The English origins and current international use of the word "forest" represents a very powerful example of a concept packed into two syllables, often separating nature from people and asserting claims on behalf of the ruler or state over forest resources to the exclusion of local people. The Magna Carta from 1215 included a promise by King Henry II to "deforest" lands recently acquired, thereby returning the control over these lands to local communities or elites. Michon et al. (2007) and Van Noordwijk et al. (2008) explored how use of the term forest and its implicit institutional and political intentions was intended to allow clear-felling and replanting as legitimate forest management practices and a means to control land and other natural resources. In contemporary times, the meaning has been inverted to ban management involving land clearing as antithetical to sound "forestry" even if the smallholder agroforestry systems, such as Indonesia's *kebun lindung* ("protective gardens"), outperform conventional forest management economically, socially, and environmentally.

Slogans can be even more insidious than single words, because they tend to pack more punch, impairing communication and enshrining fuzzy thinking. "No forest, no river," "No river, no water," "Forests prevent floods"—these slogans are commonly used in public discourse and purport to be based on sound science. Across the tropics, they have been used to justify evictions from "critical" watersheds. They also have been marshaled to justify hundreds of millions of dollars in public expenditure for reforestation under the control of forestry departments. The condensed representation of "knowledge" used in these slogans has been very powerful in the politics of control over the landscape and its resources in many countries.

An inclusive, participatory approach to developing conceptual frameworks may be an effective means of revealing the fallacies and logical errors in powerfully ambiguous words and falsely concrete slogans.

3.3 How to use conceptual frameworks to focus your assessment

Section's take-home messages

- Conceptual frameworks can provide greater focus on key issues and relationships in assessment efforts; they also can serve a useful role in synthesis and cross-site comparisons.
- The MA conceptual framework has in some ways emerged as a dominant framework for ecosystem assessment, but there is no unified theory on the creation of conceptual frameworks. Still, the structure and elements of the MA conceptual framework and related examples may be useful starting points in developing a framework.
- The MA conceptual framework (or any framework) should not be adapted for use in a new context without understanding its original purpose and deciding whether and how to adapt or translate the diagrams used to summarize its content.
- Much of the really useful bit of the approach to conceptual frameworks in the MA was not derived from a four-box diagram but from the higher-level idea linking human

well-being and ecosystem services—that vagueness opened opportunities for adaptation.
- The structure and elements of a conceptual framework cannot be comprehensive. They need to focus on the most salient issues for users of a particular ecosystem assessment.
- The structure and elements of a conceptual framework also are the foundation for identification, prioritization, and development of appropriate indicators for conditions and trends in ecosystems.
- It is important to try to keep the framework simple, paying particular attention to possible thresholds and turning points.
- Assessment teams face important choices in seeking balance between "top down" approaches to developing conceptual frameworks for synthesis and comparison and "bottom up" approaches in which collaborative development of a framework can be a means to engage stakeholders to ensure usefulness and legitimacy and thereby create a sense of ownership of the assessment process.

A conceptual framework, in the broadest sense, is a tool for organizing ways of thinking about the subject at hand: "A well-designed framework for either assessment or action provides a logical structure for evaluating the system, ensures that the essential components of the system are addressed as well as the relationships among those components, gives appropriate weight to the different components of the system, and highlights important assumptions and gaps in understanding" (MA 2003: 34). For ecosystem assessment, the "subject at hand" is the relationship between people and nature or, more specifically, between human well-being and ecosystem services. Of particular interest is how those relationships may be changing. There are many examples of conceptual frameworks but no unifying theory. And, in practice, there is a huge range in the use of the term. A conceptual framework could be a formal model, a sketch on the back of a used envelope, or a figure produced through collaborative effort. It could be accessible or off putting, could clarify or obscure, and could engage others' views or simply presume them.

A conceptual framework does not necessarily have to be represented with a diagram, but often a visual representation increases the effectiveness in conveying the ideas and elements of the framework to most audiences, including users and members of the assessment team. Done well, graphical representations can help clarify and pinpoint key issues within webs of complex relationships; they can help bring differences into focus and also aid in developing a shared terminology and data protocols that can facilitate interdisciplinary and cross-site collaboration. The framework, then, is an important—a nearly indispensable—tool for design of multistakeholder, multidisciplinary, multiscale, multisite assessments.

Use of any type of figure in clarifying thinking must be balanced against the risks of rigidity in approach and other pitfalls. On the one hand, it is well established that overly narrow analyses of environmental issues can be highly misleading (Baumol and Oates 1988). Neglecting real complexities when framing problems can lead to overly simplistic solutions that are not relevant or applicable to the intricacies and difficulties of real situations. Recognizing the complexity, interconnections, and trade-offs involved is critical. However, there is an equal risk of lack of focus and, consequently, being overwhelmed and ultimately paralyzed by detail. Conceptual frameworks can be an antidote to this problem of "everything depends on everything and everyone." But this also means that compromises are inevitable and that there is no single "best" representation.

3.3.1 Overview of the structure and main elements of the MA conceptual framework

There are many approaches to thinking through and structuring graphic conceptual frameworks (see Boxes 3.3, 3.4, and 3.5). Any graphic representation of a conceptual framework is

> "simultaneously helpful and problematic, [has] an ad hoc flavor, and should be viewed as a heuristic device. Box and arrow frameworks reflect the infancy of theoretical studies, and were designed to facilitate the quest for general principles and integrated theories." (Lambin et al. 2006:6)

The particular structure used in the global portion of the Millennium Ecosystem Assessment (see Figures 3.1 and 3.2) had its origins in the Drivers-Pressures-States-Impacts-Responses (DPSIR) framework (Smeets and Weterings 1999, Pirrone et al. 2005 and, for a critical view, see Svarstad et al. 2008). A simpler version of this, called Pressure State Response (PSR), is among the frameworks most frequently used by international development organizations (Spreng et al. 1996). The foremost difference between PSR and DPSIR is that the latter identifies drivers explicitly and introduces the impact category, which helps highlight the most salient causal relationships. Both DPSIR and PSR are linear frameworks, examining how changes in pressures affect systems but not including the full loop to consider how changes in the system then feed back to affect pressures. The MA framework differs in fundamental ways from these in that it incorporates both this feedback loop and multiple temporal and spatial scales.

The foundation of the application of DPSIR is a comparison of reference conditions (or baselines) in ecosystem services with the same or similar systems under increasing degrees of human intervention and disturbance. Historically, biologists and ecologists have sought the general rules governing natural systems by studying an idealized notion of nature in its most "pristine" state. Understanding these rules and drawing comparisons between "pristine" natural areas (reference conditions) and areas with varying degrees of human intervention is the main method used to gauge the type and extent of impacts of human activities on ecosystems (Spreng and Wils 1996). In contrast, the MA conceptual framework does not assume any "natural" state for the system and instead treats the system condition as a dynamic response to changes in drivers.

Box 3.3. Other systems approaches to thinking and graphic representation

DPSIR is closely related to systems analysis and hierarchy theory, both of which can provide additional insight into multiscale analysis. In these approaches, a "system" is perceived as a whole, with elements intrinsically related because they consistently affect one another over time and "operate towards a common purpose" (Senge et al. 1994:90). "Hierarchical systems" often are described as a nested set of subsystems, including the key feedback loops affecting the nature of the subsystem interactions (Simon 1962).

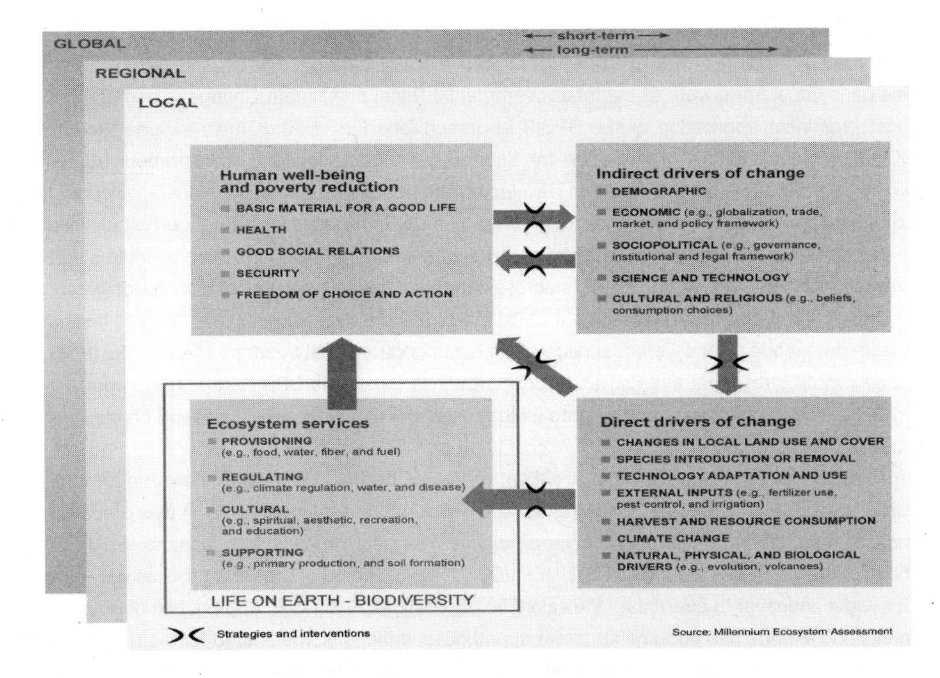

Figure 3.1. MA conceptual framework.
Source: MA 2005a

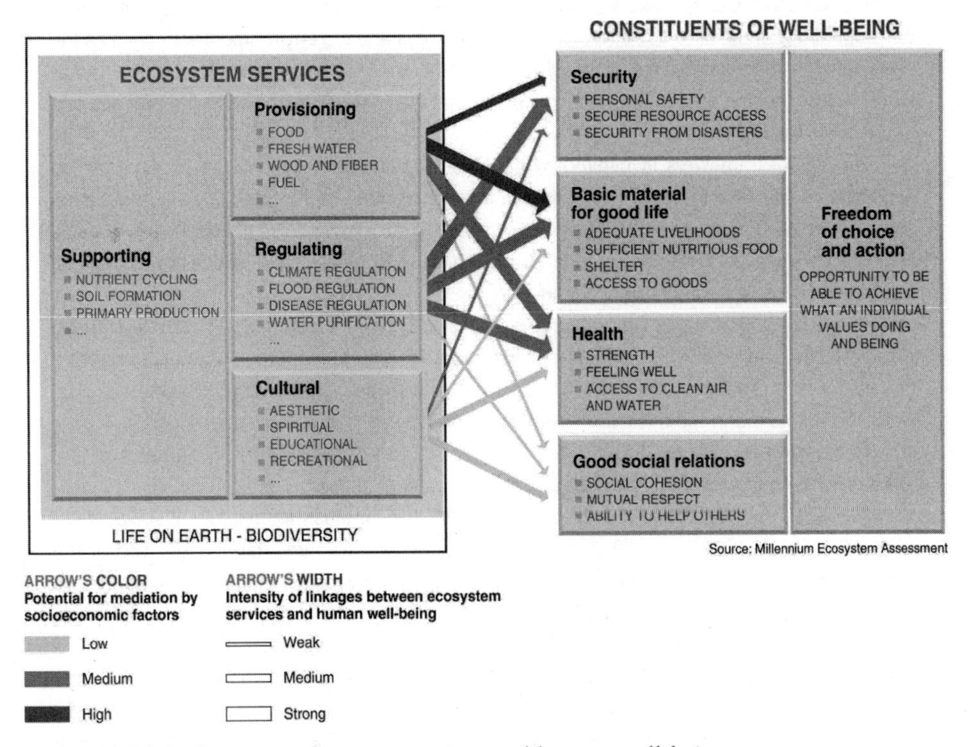

Figure 3.2. Links between ecosystem services and human well-being.
Source: MA 2005a

Box 3.4. Comparing different uses of DPSIR frameworks

The conceptual framework of the Intergovernmental Panel on Climate Change is perhaps the most prominent application of the DPSIR approach (see Figure A). (Others include the UN SCOPE project and the Organisation for Economic Co-operation and Development framework.) The IPCC has been a leader in developing the notions of differential vulnerabilities and adaptive capacities mediating impacts and has usefully introduced some new broad classes of response options (mitigation, adaptation). But whereas the IPCC focuses on a single direct driver (climate change) and on impacts across a range of specific sectors (energy, food, water), the Millennium Ecosystem Assessment considered multiple direct drivers and focused on the effects on a range of ecosystem services. The basic contrasts between the MA and the IPCC framework illustrate that even when both approaches follow the DPSIR structure, important choices need to be made about what to include as major elements, which depend on the purpose of the assessment.

The *Global Environment Outlook* (GEO), produced by the U.N. Environment Programme (UNEP) since 1997, uses the DPSIR framework. The fourth assessment (GEO-4) was prepared through an intergovernmental and scientifically independent process with features similar to those of the IPCC, MA, and IAASTD (UNEP 2007). This evolution of the GEO process was part of a wider endeavor to strengthen the scientific base of UNEP. Different approaches of governments and experts had become apparent during consultations: some argued for the use of the DPSIR conceptual framework used in previous GEOs and others argued for the MA framework (UNEP 2004).

This dilemma was tackled in GEO-4 by evolving the DPSIR framework to include concepts from the MA and IPCC frameworks. Framing an assessment of how environment contributes to development was the aim. The interface between people and the environment was seen as the carrying concept for such a frame (see Figure B). As in the MA, the development of the framework created tension among authors—a tension that was not fully resolved in the process due to time constraints.

The GEO-4 framework is inspired by the MA, but because of its broader scope it places ecosystem services and human well-being in a wider context. It prompted GEO-4 to question some of the assumptions in the MA. A minor but illustrative conceptual difference is that the MA considers climate change as a direct driver, while in GEO-4 it is seen as one among several interacting forms of environmental change.

Compared with the MA, GEO-4 expands the environmental factors that determine human well-being beyond ecosystem services. Nonecosystem natural resources and the stress that the environment imposes on society (in the form of diseases, pests, radiation, and hazards) were added to expand the basis for assessing trade-offs.

A core development in GEO-4 is its acknowledgement of the fact that environmental factors are interacting with demographic, social, and material factors in determining human well-being. A variation of the framework presented in GEO-4 illustrates the dual role of social and economic sectors (see Figure C). The sectors are driving environmental change, and at the same time they are instrumental in shaping the way the environment affects people. Again, this was considered important in order to assess real life trade-offs. But even more so, it was needed to allow in-depth analysis of people's vulnerability to environmental change.

The GEO-4 conceptual framework condenses a number of ideas into one figure. This approach illustrates the risk of overloading a framework, but it also demonstrates how well-known concepts from different realms can be combined to give new insights. In this respect it

(continued)

Box 3.4. continued

is worth noting the attempt to incorporate the concept of material, human, social, and natural capital into the equation. Also, it is worth noting that the approach to "responses" echoes, albeit broadens, the approach in the IPCC framework. In GEO-4, "responses" are seen as "formal and informal adaptation to, and mitigation of, environmental change" (UNEP 2007: figure 1, p. xxii).

Finally, the GEO-4 framework demonstrates that the concepts introduced in the MA can further evolve and combine with other approaches such as the DPSIR framework. New assessments, whether global or subglobal in nature, should continue to challenge the current conceptual understanding of the complex chains of cause and effect taking place in space and time that characterize the interactions between people and the environment.

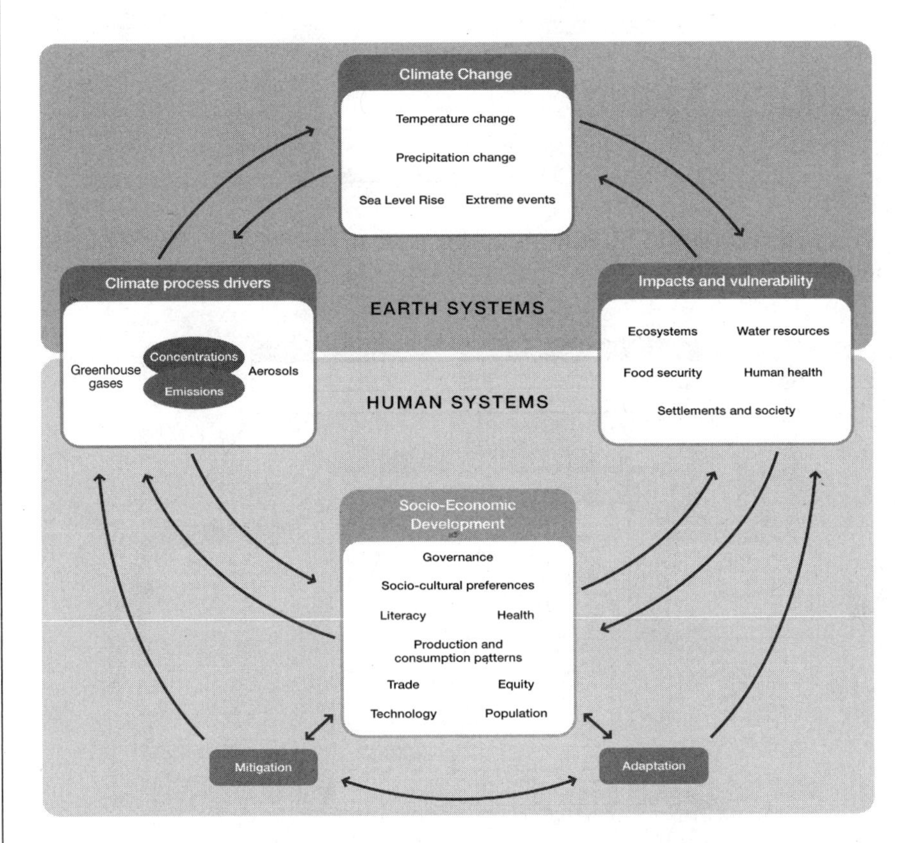

Figure A Box 3.4. Conceptual framework of the Intergovernmental Panel on Climate Change.
Source: IPCC 2007

(continued)

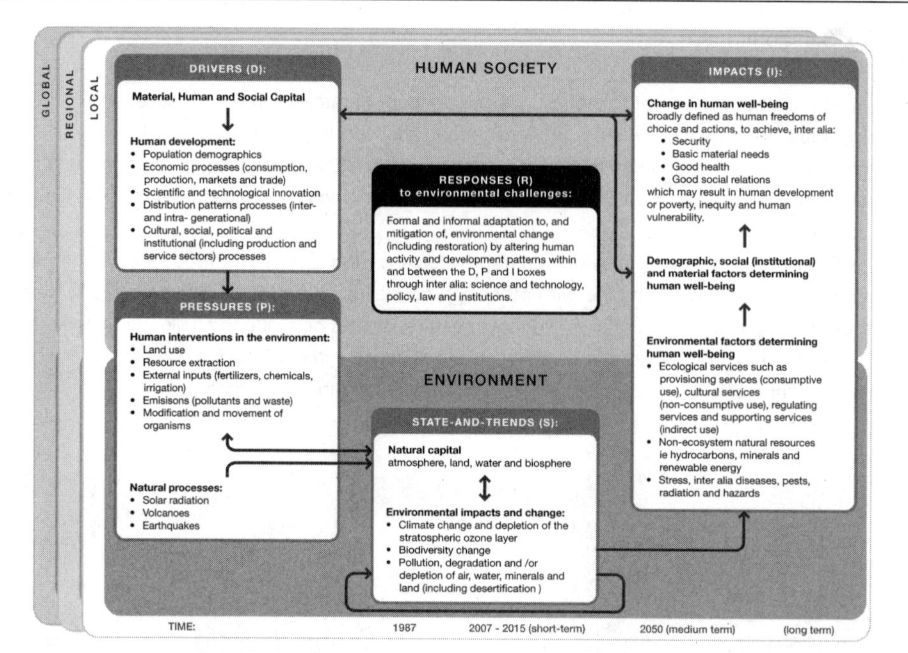

Figure B Box 3.4. DPSIR framework used in GEO-4.
Source: UNEP 2007

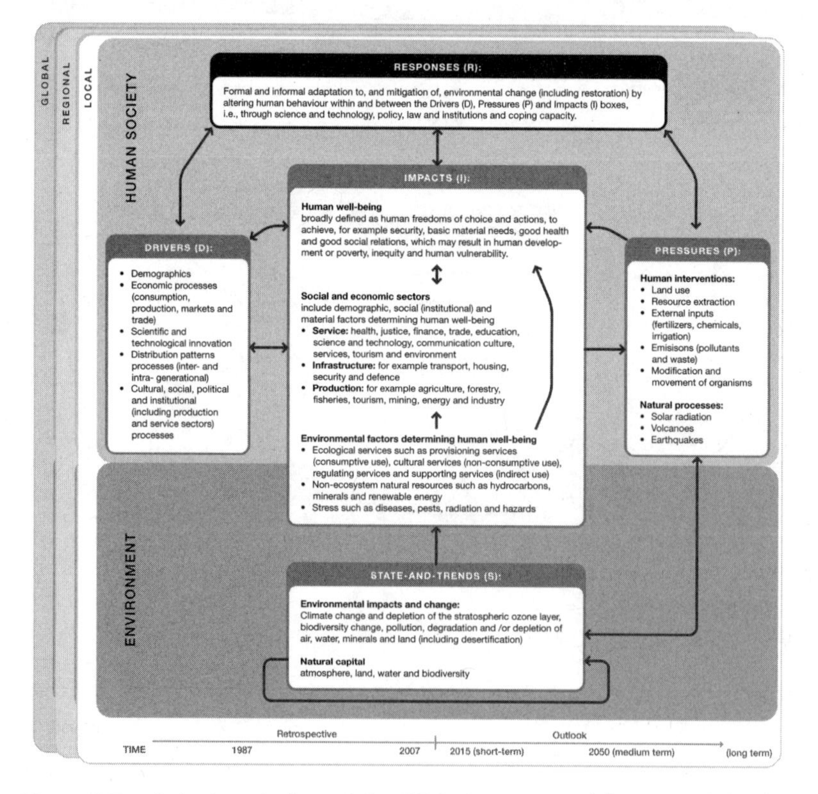

Figure C Box 3.4 . A variation of the GEO-4 conceptual framework highlighting the dual role of the social and economic sectors.
Source: UNEP 2007

(continued)

Box 3.5. Multiple conceptual frameworks for dryland systems

It is fair to conclude that the processes of degradation of dryland systems, sometimes called desertification and "defined as a persistent decrease in provisioning of ecosystem services" (MA 2005a:645), remain poorly documented at the global scale (Lambin et al. 2003). To better frame the assessment of dryland systems under the U.N. Convention to Combat Desertification (UNCCD), the European Commission (EC), through the Institute of Environment and Sustainability of the Joint Research Centre, started a process in 2008 that illustrates the role of multiple conceptual frameworks in the absence of a single, synthetic framework. The overall objective is to evaluate and integrate strategic indicators and benchmarks for a comprehensive assessment in a multistage process.

First, existing indicator systems and conceptual frameworks are critically reviewed from the perspective of the Dahlem Desertification Paradigm (DDP). While a paradigm, in general, relates to a set of practices that define scientific approaches during a particular period of time (Kuhn 1962), the DDP attempts to capture the multitude of biophysical and socioeconomic interrelationships within dryland systems from the perspective of human-environmental system dynamics (Newell et al. 2005). There are nine assertions and implications arising from the DDP (Reynolds and Stafford Smith 2002, Reynolds et al. 2007), so that in the end the paradigm might turn into "a single, synthetic framework . . . (that) is testable, which ensures that it can be revised and improved" (Lambin et al. 2007:336). These assertions are as follows (with implications not detailed here):

- Desertification always involves human and environmental drivers.
- "Slow" variables are critical determinants of system dynamics.
- Thresholds are crucial and may change over time.
- Costs of intervention rise nonlinearly with increasing degradation.
- Desertification is a regionally emergent property of local degradation.
- Coupled human–environment systems change over time.
- Development of appropriate local environmental knowledge must be accelerated.
- Systems are hierarchically nested (so, manage the hierarchy).
- A limited suite of processes and variables at any scale makes the problem tractable.

Second, indicators are held against existing conceptual frameworks to test and illustrate their potential for application. The EC initiative provides several examples of frameworks with relevance for dryland systems; two of them are illustrated here only to demonstrate their relevance under the DDP:

- *Human Ecosystem Model:* Developed at the University of Idaho in the United States, this model has the character of an organizing principle to design "a coherent system of biophysical and social factors capable of adaptation and sustainability over time." The flow and use of a limited set of critical resources are seen to be regulated by the social system, creating the so-called human ecosystem (see Figure A) (Machlin and McKendry 2005). The model is multiscale and hierarchically nested, and it is considered most useful for predicting and evaluating cascading and nonlinear effects; it is also able to synthesize a large range of theory, method, and evidence, including dryland systems.

(continued)

Box 3.5. (continued)

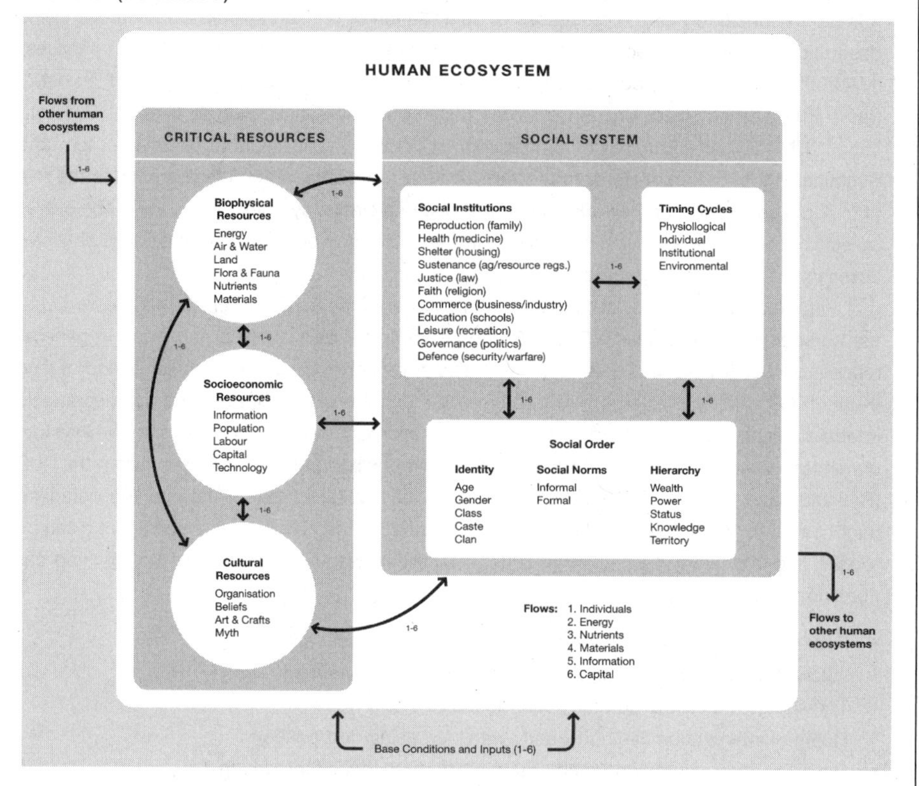

Figure A Box 3.5. Human Ecosystem Model.
Source: Machlin and McKendry 2005

- *Adaptive Cycle Model:* Developed by the Resilience Alliance, a collaboration of scientists and practitioners to explore the dynamics of socioecological systems (Folke et al. 2002), the model is a "tool of thought" to focus on the interaction of processes of destruction and reorganization in complex adaptive systems (thus moving beyond traditional notions of stable state and succession). It identifies four distinct phases with different rates of change; the two newly added functions are rapid transitions that occur from "collapse" (or release) to reorganization (the so-called Omega>Alpha backloop), while the so-called foreloop (r>K) is modeled as a slow, incremental process of growth and accumulation (see Figure B). Adaptive cycles are seen to be nested in hierarchies across time and space, representing a panarchy (Gunderson and Holling 2002). This implies that the results obtained from drylands would not trigger cascading instabilities of the whole system because of the stabilizing nature of nested hierarchies.

(continued)

Box 3.5. (continued)

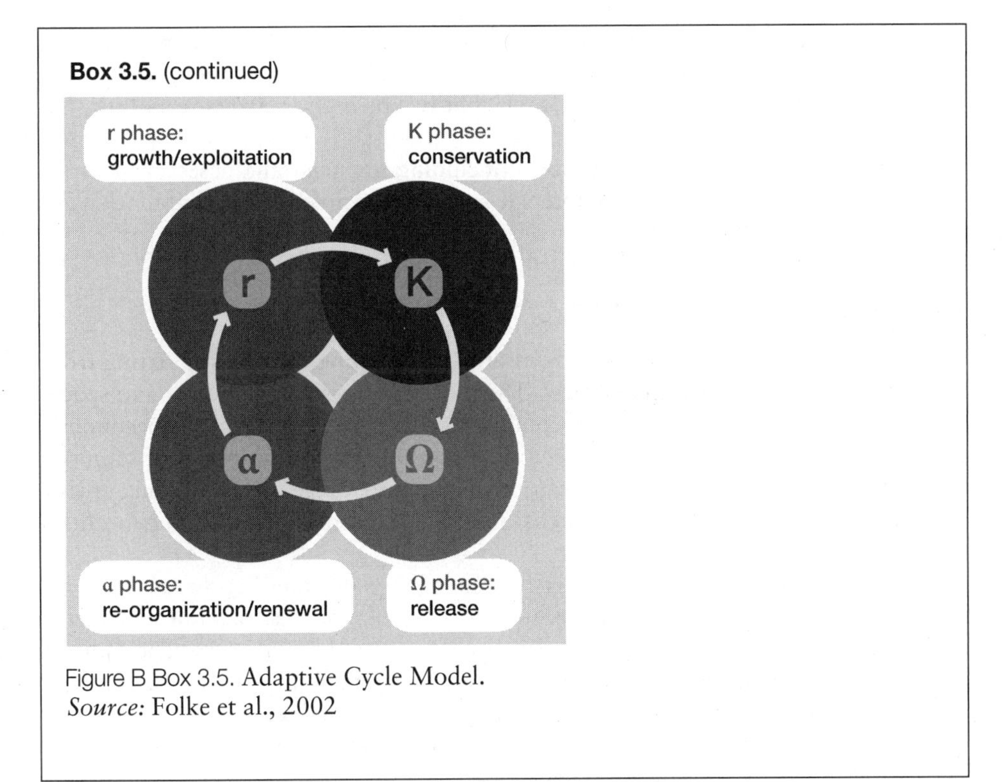

Figure B Box 3.5. Adaptive Cycle Model.
Source: Folke et al., 2002

This section briefly considers each of the main components of the MA conceptual framework:

- *Indirect drivers* (a subset of the "pressures" in a PSR framework); also sometimes called underlying drivers.
- *Direct drivers*, which, along with indirect drivers, constitute the "pressures" in a PSR framework. These also are called "driving forces," "process variables," or "control variables" in other frameworks. Indicators in this category gauge a process that will influence conditions and trends (state variables).
- *Conditions and trends* in biodiversity, ecosystem services, and human well-being; known as "state variables" in PSR frameworks.
- *Impact variables*, showing cause-and-effect relationships linking different elements. Impacts are symbolized by the arrows in Figure 3.1 and are elaborated in Figure 3.2 for the specific cause-and-effect relationships linking various ecosystem services and elements of human well-being. In a PSR framework these are unidirectional, whereas in the MA framework bidirectional impacts are allowed. Although there is an element of bidirectionality in many relationships, double-headed arrows should be used sparingly and reserved for the most salient feedbacks in a conceptual framework.
- *Response variables*, which are efforts by people (individually or collectively) to reduce negative impacts and enhance positive impacts in order to attain more desirable levels in the pressure and state variables. Responses are symbolized by the symbols marked "strategies and interventions" in Figure 3.1. Assessment

of these strategies and interventions is the focus of Chapter 6. Participatory appraisal, policy analysis, cost–benefit techniques, impact assessment, and outcome mapping are a few of the tools that can be used to assess responses.

Conditions and trends in state variables (including impacts) and assessment of responses are treated in detail in Chapters 4 and 6, respectively. This chapter devotes additional attention to drivers.

3.3.2 Drivers of ecosystem change

Ecosystem change is always caused by multiple interacting factors originating from different levels of organization (scales). The mix of drivers varies in time and space; some common changes result from a combination of biophysical and socioeconomic drivers that work gradually (e.g., declining infant mortality, the spread of salinity), while others happen precipitously (e.g., drought, hurricane, war, economic crisis). The specific abiotic and biotic factors that merit most attention vary among localities and regions (Lambin et al. 2003).

As noted earlier, factors causing ecosystem change do so either directly or indirectly. A direct driver unequivocally influences ecosystem processes, while an indirect driver operates more diffusely, by altering one or more direct drivers. The indirect drivers are underlying (root) causes that are formed by a complex of social, political, economic, demographic, technological, and cultural variables. Collectively, these factors influence the level of production and consumption of ecosystem services. The causal linkage is almost always mediated by other factors, thereby complicating statements of causality or attempts to establish the proportionality of various contributors of change.

Meta analyses suggest that the five most important groups of indirect drivers to consider in developing a conceptual framework for ecosystem assessment are:

- Population change (demographic drivers);
- Change in economic activity (economic drivers);
- Sociopolitical drivers;
- Cultural (and religious) drivers; and
- Technological change (science and technology).

Important direct drivers to consider in a conceptual framework for ecosystem assessment include:

- Habitat changes (driven through land use/cover change, physical modification of rivers, or water withdrawal from rivers);
- Overexploitation;
- Invasive alien species;
- Pollution; and
- Climate change.

Taking this list one step further, direct drivers of land use change and land cover change (e.g., deforestation, desertification, human settlement) are activities such as logging, cropland expansion, road building, and other types of infrastructure development (Geist and Lambin 2002, Geist and Lambin 2004, MA 2005a:73–76, Nelson et al. 2006).

The various individual causes or drivers of ecosystem change, such as those just listed, interact directly, are linked via feedback, and thus often have synergistic effects. Conceptual efforts to frame these interactions relate to pathways, trajectories, causal clusters, or "syndromes." Common to such conceptual frameworks is that not all causes of change and not all levels of organization are equally important, that certain driver combinations appear repeatedly, and that a limited suite of processes and variables at any scale makes the problem tractable (MA 2003, Lambin et al. 2003, MA 2005a:73–76). At least two broad strands of treatment of drivers in conceptual frameworks can be distinguished: the concept of clusters, pathways, or trajectories, and the syndromes concept.

The clusters, pathways, or trajectories concept was developed within the Land-Use/Cover Change (LUCC) Programme (Lambin et al. 2006). This framework helped the MA distinguish indirect drivers ("underlying driving forces") and direct drivers ("proximate causes") (MA 2003, Reid et al. 2006). Summarizing from a large number of empirical–analytical cases studies, it has been found that land change is driven by a combination of a limited number of fundamental high-level causes (resource scarcity, changing opportunities, policy interventions, vulnerability, sociocultural factors) that combine direct and indirect drivers in a typical situation, making a difference between slow and fast variables (see Table 3.1) (Lambin et al. 2003:224).

In tropical forest and (sub)tropical dryland systems, frequent and recurrent ("robust") driver combinations can be distinguished (Geist et al. 2006). In tropical forests, a contemporary pattern stems from the necessity for road construction that is associated with wood extraction or agricultural expansion. Such expansions are mostly driven by policy and institutional factors, but they also involve economic and cultural drivers (e.g., frontier mentality), with variations of the pattern existing across time and space (Geist and Lambin 2002, MA 2005a:585–621). In (sub) tropical dryland ecosystems, a recurrent pattern of causal interactions stems from the necessity for water-related infrastructure that is associated with the expansion of irrigated croplands and pastures, mostly driven by policy, economic, and technological factors (Geist and Lambin 2004, MA 2005a:623–62). Again, the pattern has been found to vary in space and time (Geist 2005).

The syndromes concept was developed at the Potsdam Institute for Climate Change in 1994–2000 as a transdisciplinary core project. The concept describes archetypical, dynamic, and coevolutionary patterns. A list of about 100 "symptoms" of change (e.g., agricultural intensification) has been reduced to six "syndromes." Borrowing from medical science, a syndrome is defined as a typical cluster of symptoms and their interrelations, and it has been found that syndromes once identified will form clusters that only weakly interact with each other. Each syndrome (Sahel, Aral Sea, Dust Bowl, overexploitation, favela, mass tourism) links processes of degradation to both changes over time and the status of state variables (Schellnhuber et al. 2002). For example, the so-called Aral Sea syndrome describes socioecological deterioration as a consequence of large-scale infrastructure projects, and the Sahel syndrome refers to the overuse of marginal agroecosystems by poor, impoverished rural populations with little or no livelihood options, thus triggering further degradation and poverty.

Table 3.1. Typology of the causes of land use change

Speed of change	Resource scarcity causing pressure of production on resources	Changing opportunities created by markets	Outside policy intervention	Loss of adaptive capacity and increased vulnerability	Changes in social organization, in resource access, and in attitudes
Slow	Natural population growth and division of land parcels Domestic life cycles that lead to changes in labor availability Loss of land productivity on sensitive areas following excessive or inappropriate use Failure to restore or to maintain protective works of environmental resources Heavy surplus extraction away from the land manager	Increase in commercialization and agro-industrialization Improvement in accessibility through road construction Changes in market prices for inputs or outputs (e.g., erosion of prices of primary production, unfavorable global or urban–rural terms of trade) Off-farm wages and employment opportunities	Economic development programs Perverse subsidies, policy-induced price distortions and fiscal incentives Frontier development (e.g., for geopolitical reasons or to promote interest groups) Poor governance and corruption Insecurity in land tenure	Impoverishment (e.g., creeping household debts, no access to credit, lack of alternative income sources, and weak buffering capacity) Breakdown of informal social security networks Dependence on external resources or on assistance Social discrimination (ethnic minorities, women, lower class people, or caste members)	Changes in institutions governing access to resources by different land managers (e.g., shift from communal to private rights, tenure, holdings, and titles) Growth of urban aspirations Breakdown of extended family Growth of individualism and materialism Lack of public education and poor information flow on the environment
Fast	Spontaneous migration, forced population displacement, refugees Decrease in land availability due to encroachment by other land uses (e.g., natural reserves or the tragedy of enclosure)	Capital investments Changes in national or global macro-economic and trade conditions that lead to changes in prices (e.g., surge in energy prices or global financial crisis) New technologies for intensification of resource use	Rapid policy changes (e.g., devaluation) Government instability War	Internal conflicts Illness (e.g., HIV) Risks associated with natural hazards (e.g., leading to a crop failure, loss of resource, or loss of productive capacity)	Loss of entitlements to environmental resources (e.g., expropriation for large-scale agriculture, large dams, forestry projects, tourism and wildlife conservation), which leads to ecological marginalization of the poor

Source: Lambin et al. 2003:224.

Syndromes reflect both expert opinion and local case study information. The approach is applied at the intermediate functional scales that reflect processes taking place from the household level up to the international level. Syndromes aim at a high level of generality in the description of mechanisms of environmental degradation rather than processes of restoration, renewal, or reorganization. As for land degradation in (sub)tropical dryland systems, several syndromes may be applied (e.g., "Sahel," "Aral Sea," and "overexploitation"). The concept and the Sahel syndrome, in particular, have been criticized because of the exclusive use of downward spirals, inherent Malthusian thinking, tautological reasoning, and the lack of actor orientation (Geist 2006).

As for decision making at the local level, a distinction can be made between exogenous drivers—that is, those outside the control of local land managers and mainly including indirect drivers (prices, markets, technology development) but also a few direct drivers (e.g., climate change)—and endogenous drivers—that is, those under the control of local managers and mainly including direct drivers (use of external inputs, species introductions and removals), with only a few indirect drivers (e.g., technology adaptation). The implication is that, conceptually, intervention measures might be targeted to underlying causes (indirect drivers) rather than proximate ones (direct drivers) and to weakening the positive feedbacks that accelerate unsustainable ecosystem change while strengthening the negative feedbacks that slow or dampen ecosystem change (see Figure 3.3). The conceptual model shows where, during the processes of LUCC, national policy is likely to have the most impacts on land use (areas indicated with a superscript 2) and where interventions will be more difficult (areas indicated with a superscript 1) (Reid et al. 2006:162).

3.3.3 Flows of ecosystem services, stocks of resources, or both as state variables?

Flows of ecosystem services per unit time are closely linked to changes in the levels (stocks) of resources needed to produce those flows: there can be little cycling of soil nutrients (a flow of fertility) without the necessary populations of soil organisms (a stock of biological resources).

The MA conceptual framework emphasizes flows of ecosystem services and does not work as well regarding the stocks of resources that are essential to sustainability. Though flows of ecosystem services may sometimes be an effective proxy for the status of the underlying natural resource base, this is not always the case. Increasing pumping of groundwater (an increasing service flow) can be (and often is) associated with aquifer depletion (a change in a stock).

Resource stocks are increasingly recognized as crucial variables for measurement, presumably because they can be used to determine how resources will persist under current or future patterns of use (Victor 1991). Sustainability assessments informed by this school of thought often emphasize maintaining capital stocks (Stern 1995), which are also described as state variables (Ludwig et al. 1997, Bell and Morse 1999). This "capital theory" grew from the extensive literature on economic growth and finite resources that thrived in the 1970s (Victor 1991). Work by the U.K. Department for International Development (DfID) is one prominent example of the capital theory approach (often called the "livelihoods approach"), which focuses on "five types of capital" (natural, physical, financial, human, and social) as a

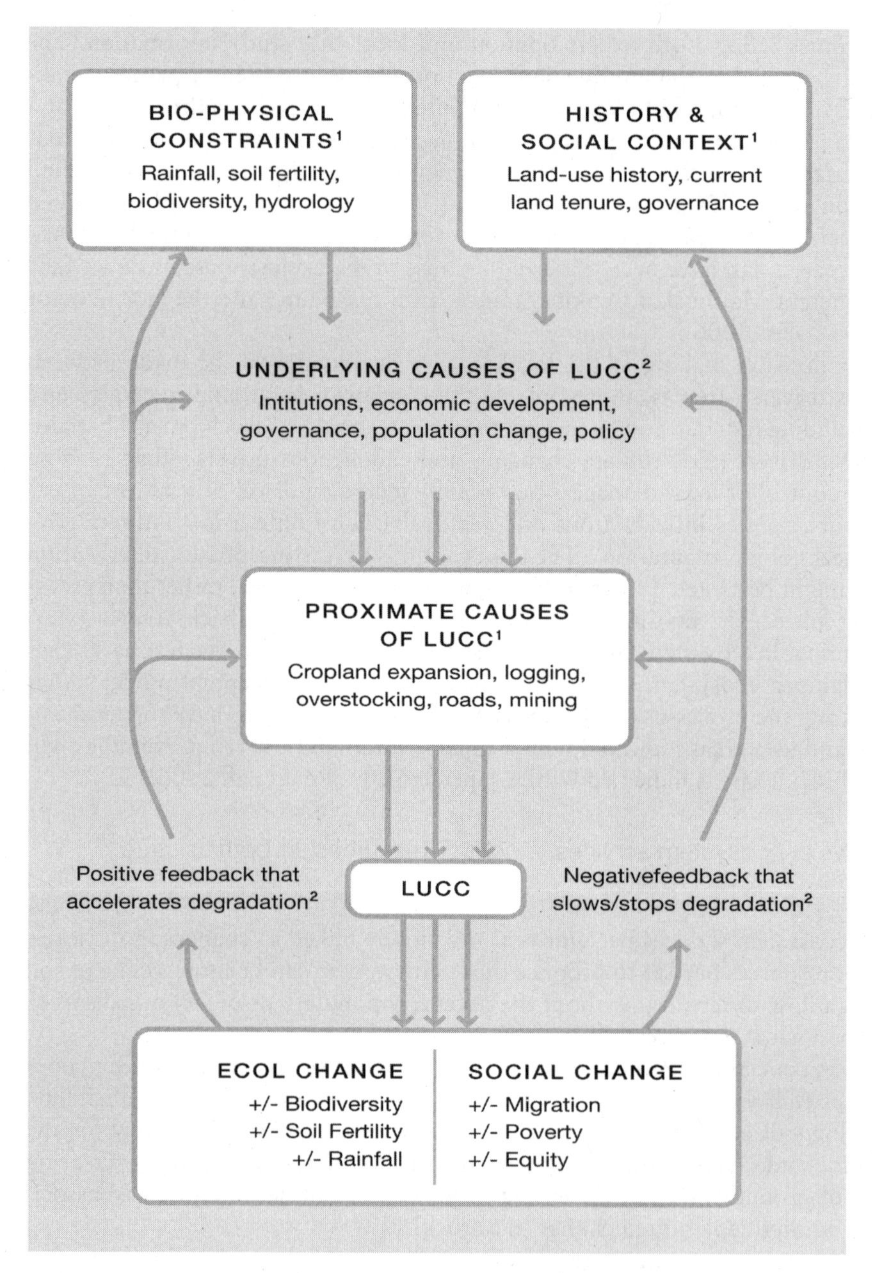

Figure 3.3. Conceptual model of land use and land cover change.

conceptual framework for planning, programming, and project design for sustainable development.

Arrow et al. (2004) have proposed a consistent and empirically tractable concept of sustainability that provides a useful framework for focusing on the resource base, interpreted broadly in terms of the stocks of the five types of capital developed by DfID. The upper right box in Figure 3.4 incorporates a list of resource stock

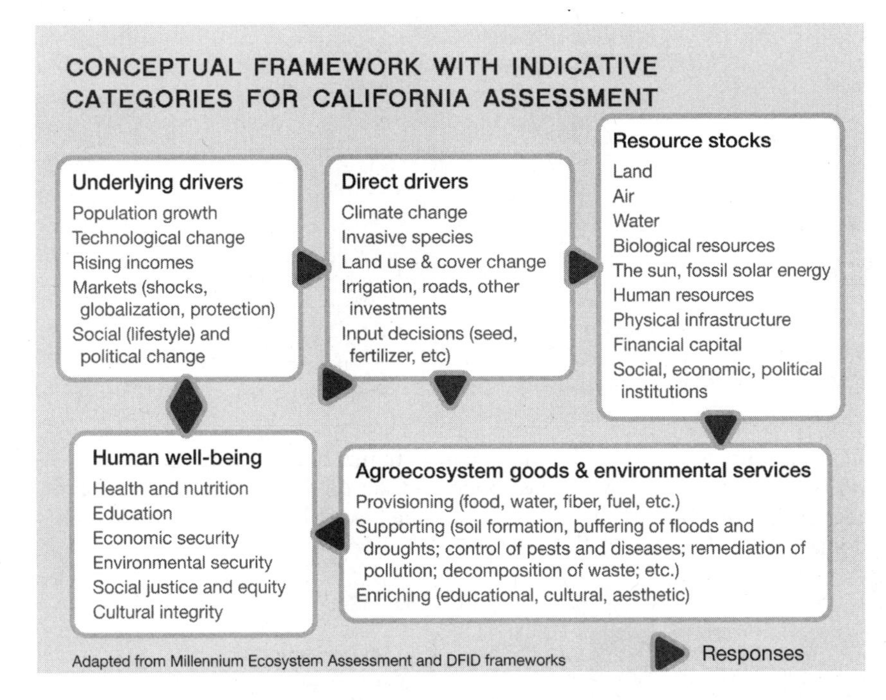

Figure 3.4. Modified conceptual framework for California Agroecosystem Assessment. *Source:* Adapted from Millennium Ecosystem Assessment and DFID frameworks

variables within a DPSIR framework adapted for agroecosystem assessment in California. Of course, adding these variables would significantly increase the measurement challenges and costs in any attempt to monitor the full range of indicators.

3.3.4 Multiple scales in space and time

DPSIR approaches have been criticized for their linearity and promotion of one-dimensional thinking about relationships among drivers and state variables. The potential danger lies in obscuring rather than revealing critical relationships and encouraging one-dimensional, quick-fix responses (Bell and Morse 1999). Similarly, linear DPSIR approaches have been deemed to be overly narrow, as they are unable to account for the background processes that determine ecosystem and environmental health (Berger and Hodge 1998). More dynamic approaches that emphasize feedbacks and multiple scales may help reduce these criticisms.

The MA framework differs from the basic DPSIR approach in two important ways. First, the MA framework (see Figure 3.1) incorporates multiple spatial scales (see Figure 3.5 for additional examples), which are not commonly included in PSR. Second, the MA incorporated feedbacks from both environmental changes and related consequences on human well-being over time. An important aspect of this second point is that feedback loops make the MA a more dynamic system, which can be described as circular or webbed, in comparison to the relatively linear DPSIR approach.

The MA framework's embrace of multiple spatial scales essentially combines DPSIR with the idea of nested spatial domains. As noted earlier, DPSIR has a great advantage over PSR in its attention to the causality of underlying impacts, and

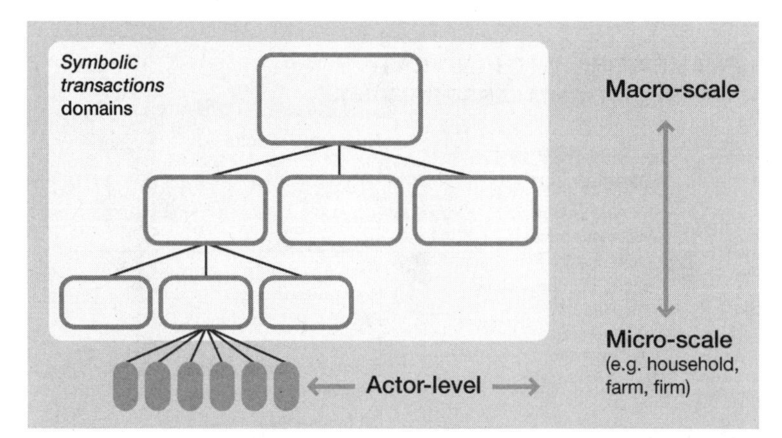

Fig. 3.5.A. Conventional nested spatial domain conceptualization as proposed by Hägerstrand. The figure shows how the symbolic transaction domains are situated and operate under each level above them. The actors at the lowest level are then left with a set of freedoms to carry out their daily activities.

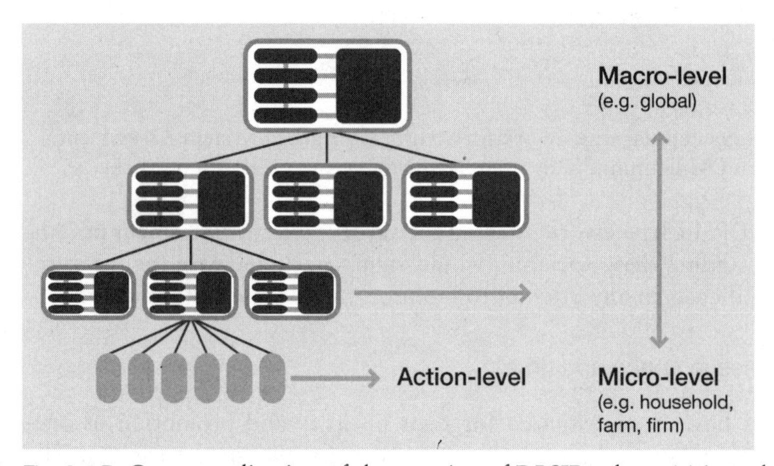

Fig. 3.5.B. Conceptualization of the merging of DPSIR scheme(s) into the system of nested spatial domains. The taxonomy allows for a focus on both actor constraints and the problem's causal linkages at each level.

Figures 3.5 A and B. Approaches to graphic depiction of spatial scale.
Source: (Ness et al. 2008)

adding more spatial scales becomes even more powerful. This approach also can be used to integrate actors who only operate in certain spatial domains, such as some official agencies (see Box 3.6).

At some point, though, there is a limit to the number of spatial and temporal scales any assessment team can handle. Since many of the differences among uses relate also to spatial or temporal scales, it makes sense to consider dividing into a multilevel collection of nested assessments, so that various issues can be analyzed at appropriate levels without being overwhelmed by the complexity of dealing at the same level of detail for the whole system (see Figure 3.6).

Box 3.6. Drivers link across scales

Summarizing a large number of case studies across various ecosytems, Geist and Lambin (2002, 2004) found that typically three to five indirect drivers underpin two to three direct drivers. These findings also shed some light on possible cross-scale dynamics. For example, for changes at tropical forest margins (Table A), it appears that local-to-global interplays are much more common than in the case of dryland systems (Table B), where local-to-national interactions are much more widespread. This insight suggests that global ownership of forest ecosystem assessment appears to be achievable, but not the global ownership of dryland ecosystem assessments. The latter might explain part of the UNCCD difficulties in achieving a comprehensive assessment and in implementing top-down measures in general (Geist et al. 2006).

Table A. Indirect drivers of tropical deforestation by scale of influence

	All factors (range)	Demographic factors*	Economic factors	Technological factors	Policy and institutional factors	Cultural or sociopolitical factors
	N=152 cases	(n=93)	(n=123)	(n=107)	(n=119)	(n=101)
Local	2–88%	88%	2%	23%	4%	16%
National	1–14%	1%	14%	3%	2%	7%
Global	0–1%	–	1%	–	–	–
Several scales: Global–local interplays	11–94%	11%	82%	74%	94%	77%

*Six cases of unspecified population pressure could not be attributed to scales.

Geist et al. 2006:64.

Table B. Indirect drivers of desertification by scale of influence

	All factors (range)	Demographic factors*	Economic factors	Technological factors	Policy and institutional factors	Cultural or sociopolitical factors	Climatic factors*
	N=132	(n=73)	(n=79)	(n=91)	(n=86)	(n=55)	(n=114)
Local	12–29%	23%	18%	29%	12%	16%	–
National	4–20%	–	13%	–	20%	4%	–
Global	4–12%	–	4%	–	6%	–	12%
Several scales: national–local interplays	29–80%	29%	66%	71%	63%	80%	60%

*Thirty-five demography-driven and 32 climate-driven cases could not be attributed to scales.

Geist et al. 2006:65.

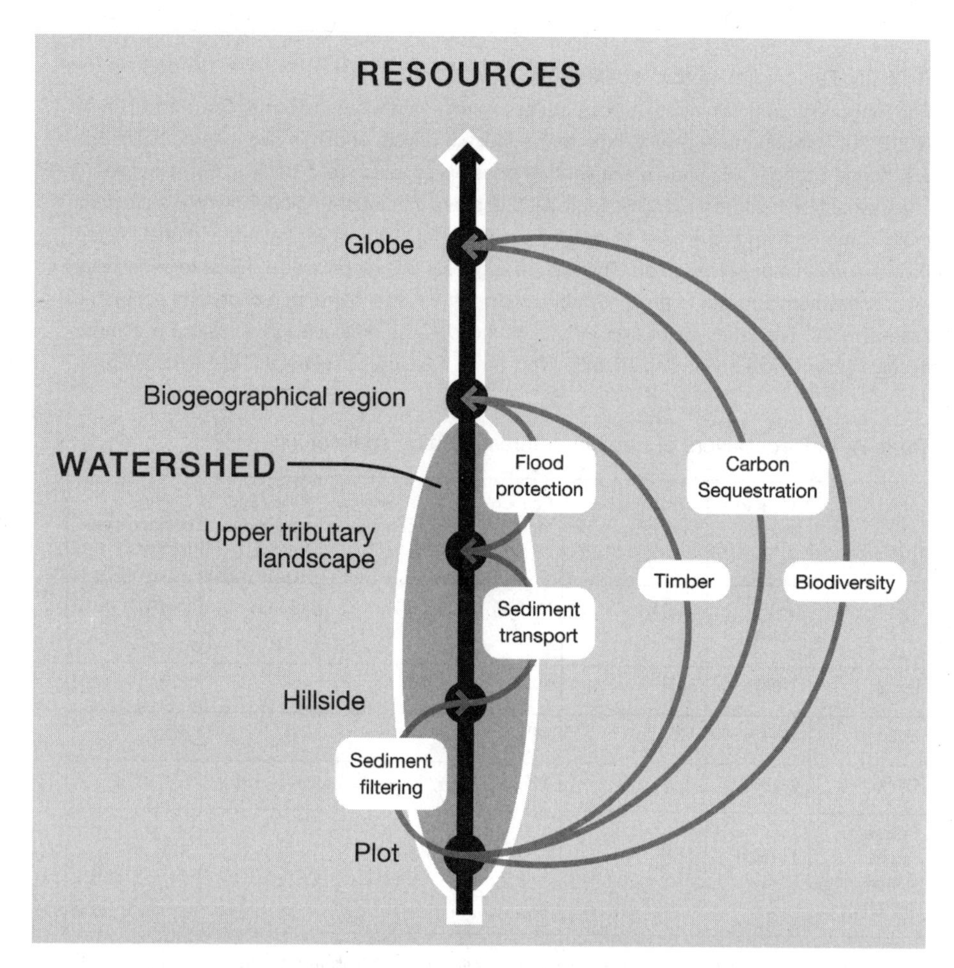

Figure 3.6. Upper tributary watersheds provide habitat for *in situ* conservation of biodiversity and ecosystem goods and services valued at other levels.
(*Source:* Lebel et al. 2008)

Multilevel approaches also may help reduce the tensions that would otherwise arise in a monolithic single-level approach (Lebel 2006). When ways of using and managing different goods and services interact, then understanding both past trends and future possibilities in quantitative, linear ways can be extremely difficult. Pluralistic devices, like developing some parallel assessments from different perspectives, may help deal with complexity.

3.3.5 Complexity and uncertainty

Complexity and uncertainty feature in every aspect of an ecosystem assessment and arise from a number of sources.

- *Incomplete and imperfect data.* Researchers never have complete information, and all data have some associated uncertainty; "reserved words" and other techniques for systematic assessment of uncertainty in quantitative and qualitative data are reviewed in Chapter 4.

- *Uncertainty about the course of future events and about the effectiveness of responses to those events.* "Pure" uncertainties about future developments can be explored using scenarios; these and related methods are presented in Chapter 5 (also see MA 2005b:229–59). Uncertainties about impact and effectiveness of responses are considered in Chapter 6.
- *Uncertainties about how complex systems work (and which parts matter most).* Lack of knowledge of key causal relationships can be dealt with by proposing alternative conceptual frameworks and treating them as competing frameworks. But when there are too many uncertainties of this sort, an assessment may be very hard to conduct. Nevertheless, it still may serve the purpose of identifying where further monitoring, observations, and research are needed (Heinz Center 2006).
- *Uncertainty about the shape of key empirical relationships.* Jared Diamond and others (e.g., Chapin et al. 2000) have emphasized the past and possible future roles of rapid, unanticipated catastrophe in social and ecological history. It is important to know when the world is getting close to the edge. But identifying which thresholds and turning points should be included in a conceptual framework for an assessment is not easy. It depends on attaining a good enough understanding of the ecosystem that, for example, alternative stable states can be identified (Resilience Alliance & Sante Fe Institute 2008). Thresholds, tipping points, discontinuities, and other nonlinearities—once acknowledged—also create challenges for representation in the textual and graphical summaries typically developed for assessments. For example, arrows drawn to indicate relationships among actors or ecosystem components may need to be completely reconfigured if a certain condition or circumstance occurs. One simple method might be to show the assumptions in two extreme cases (alternative stable states if these exist) as adjacent panels (see Figure 3.7).
- *Adaptive management, social learning, and muddling through.* Finally, a common source of uncertainty is the presence of multiple interests and kinds of users, who consequently have different perceptions about what the valued or relevant set of "ecosystems goods and services" are or what should be included in a conceptual framework and given priority for assessment (Lebel et al. 2008, Lebel and Garden 2008). This is illustrated in the case of upper tributary watersheds in Figure 3.6 and is considered further in the next section. Adaptive management and evolutionary approaches focus explicitly on social learning processes to address uncertainty. The theory of adaptive management rests on the recognition that humans often do not have the information or capacity necessary to manage ecosystems, which are dynamic, self-organizing systems that undergo cycles of instability and resilience (Walters and Holling 1990). Adaptive management has been described as a partner in scientific discovery that surveys the environment of interest and provides updated information and understanding that can be used to guide decision making. In environments where uncertainty is high, there are few reliable answers, but potential solutions can be explored through experimentation. Lee describes adaptive management approaches "as experiments that probe the responses of ecosystems as people's behavior in them changes" (Lee 1999: 1). Prominent examples of this approach include logical frameworks, strategic mapping, neighborhood sustainability

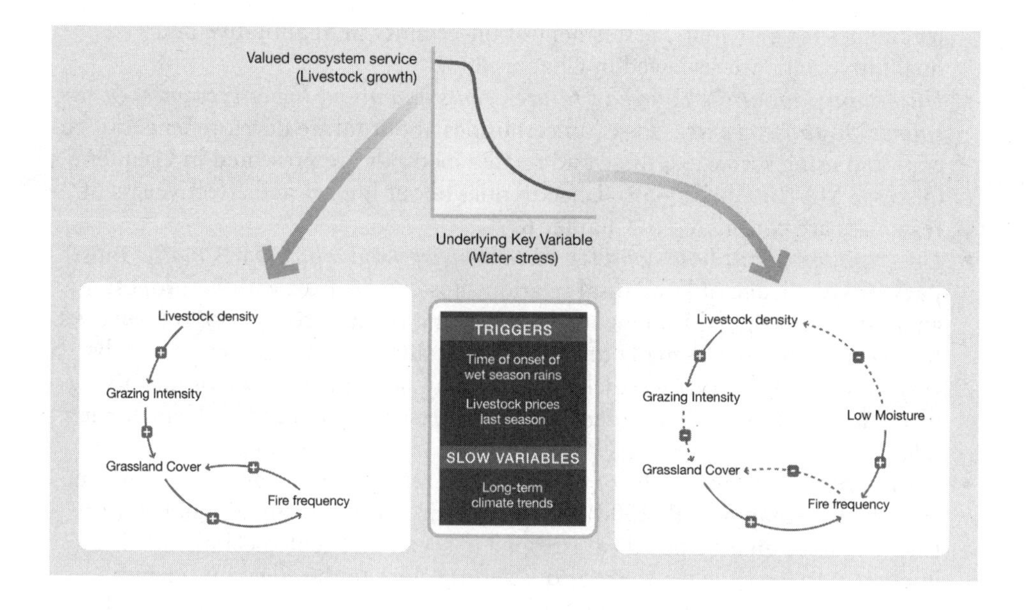

Figure 3.7. One way to represent thresholds as part of graphic conceptual framework for an ecosystem assessment.
Source: Lebel 2006

indicators (Meter 1999), integrated assessments (Kasemir 1999, Bell and Morse 2003), and systematic sustainability analysis (Bell and Morse 2003).

Many of these challenges of complexity and uncertainty are best addressed with the participation not just of the scientific and policy users of an assessment but also the users of the ecosystem themselves. Multistakeholder processes that encourage informed input and deliberation from a wide variety of sources may be needed to negotiate an acceptable initial framework to guide the assessment (Dore 2007, Warner 2006).

3.3.6 Unpacking the MA conceptual framework

Experience with the MA sub-global assessments showed it is not always easy to balance "top down" and "bottom up" tensions between assessment teams and users operating at different scales and from different perspectives. Tensions across scales and among different groups are inevitable. From the global perspective, the process of constructing a conceptual framework for ecosystem assessment within the framework of multilateral environmental agreements cannot help but obscure cultural differences and political conflict. There also is an important distinction between the treatment of complexity and uncertainty in a necessarily abstract conceptual framework, as just described, and the messy political complexity that is among the most salient considerations for local people and political leaders.

In a world of great complexity and imbalances in political power, it is not surprising that difficulties can be encountered by teams working at local and national scales. Moreover, regardless of the scales in question, it is important to note that some situations cannot be "assessed" by concepts represented in a single, global conceptual framework. The two basic MA conceptual framework diagrams (Figures 3.1 and 3.2) are commonly accompanied by a third one (see Figure 3.8), which shows how the framework has been adapted in order to make sense to a group of indigenous people in Peru. The juxtaposition of these three diagrams gives the impression that the main MA figures represent, in graphic form, a single global and scientific knowledge system, while Figure 3.8 represents one of many local and indigenous knowledge systems. This figure does show one attempt to adapt a conceptual framework to local conditions, but it does not provide any reason to believe that the result always should look like this.

Participants in the process of developing and debating the MA's conceptual framework know that it is essentially a work of compromise, a negotiated agreement to use certain words and phrases in preference to others and to adopt some broad definitions and assumptions that constitute the lowest common denominator of academic debate about the relationship between "people" and "nature," between "the economy" and "the environment," or between "human well-being" and "ecosystems." Without this common language, an assessment team would not be able to produce a coherent and persuasive assessment of what is happening to these relationships. And when a team is faced with the task of rephrasing or translating this common language for use in a social, cultural, and political environment, they need to think through how words and phrases such as "biodiversity," "ecosystem services," and "human well-being" are treated within the MA's conceptual framework and consider whether adaptations are necessary.

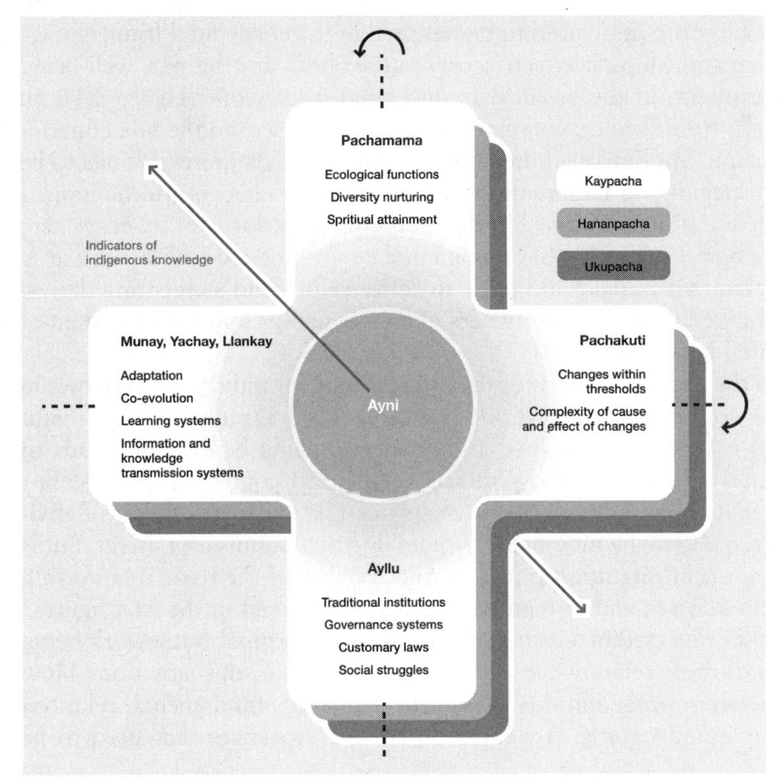

Figure 3.8. Local adaptation of the MA Conceptual Framework for the Peru sub-global assessment. *Source:* MA 2005b

This section includes a sample of some contentious issues that may merit particular attention by an assessment team that is working with assessment users to adapt or create a conceptual framework. It is worth taking some time to explore these questions (and delving into others as well) to ensure that the conceptual framework develops in a way that not only is useful to the stakeholders but also engages them in the assessment process from its outset. Doing so not only should help lend insight into the goals and origins of the MA. It also may help the assessment team clarify aspects that fit the circumstances and key elements that may need to be adjusted to reflect local conditions or gaps that need to be filled.

For example, the definition and measurement of human well-being is an intensely political issue. What types of people and human behavior are most important in the assessment being undertaken? How do different peoples' value systems affect the depiction of human well-being in the conceptual framework? How can different peoples' knowledge systems be included? How does biodiversity figure in these considerations of human well-being?

In many countries today, the central challenge for ecosystem management and human well-being is the relationship between poverty (or inequality) and the environment. How will these relationships be handled in the conceptual framework? Is it important to allow for the sort of poverty–environment relationship described in the Brundtland Report, where "poor people are forced to overuse environmental resources to survive from day to day, and their impoverishment of their environment further impoverishes them, making their survival ever more difficult and uncertain" (WCED 1987:27)? Of course, affluence may also have this kind of influence. In either case, there will be many other things that cause rich people or poor people to behave in certain ways that may need to be reflected in the conceptual framework.

This web of relationships between ecosystem services and human well-being deserves careful thought. In the basic MA conceptual framework (Figure 3.1), an arrow leads upward from the box containing ecosystem services to the box containing human well-being. But unlike all the other arrows in this diagram, it looks to be immune from "strategies and interventions." So why might this control be limited or entirely absent in the one relationship that is the primary focus of an ecosystem assessment? One answer might be that human and environmental well-being are so intimately connected that it makes no sense to conceive of them as separate things. That answer might well appeal to the owners of an indigenous knowledge system of the kind represented in Figure 3.8.

Consider also the appropriateness of the impacts that are indicated by arrows in the two basic MA conceptual framework figures (Figures 3.1 and 3.2). Why might the only arrow to escape from the box of human well-being be one that leads to indirect drivers, and why is this arrow subject to political control? It may be fair enough to argue that the supply of ecosystem services, or the social effects of environmental change, can only be modified by something that counts as a driver. But is the poverty–environment relationship just another version of the basic relationship between ecosystem services and human well-being, as depicted in the MA figures? Or does it need to be depicted in a different way in the conceptual framework being prepared? Which of these relationships are most important in this situation? How would different groups engaged in the assessment depict this fundamental relationship between people and nature? Are there different perspectives that need to be addressed?

3.4 How to use conceptual frameworks to engage users

Section's take-home messages

- Assessment processes are inherently political, involving disparities in power and authority and conflicting interests.
- Various pragmatic approaches vis-à-vis the global conceptual framework are valid—from adaptation to independence and including the use of multiple frameworks.
- The people who variously are (or are not) informed, consulted, and involved in the creation of the conceptual framework for ecosystem assessment and the ways in which their knowledge and expertise are valued (or not) will in many ways govern the entire assessment process.
- There will be significant disparities in power and authority among groups that have important stakes in the human–environment interactions at the heart of an ecosystem assessment.

Conceptual frameworks are used by their proponents in several different ways, and these can be mapped against the top-down to bottom-up spectrum of ways to go about framing an ecosystem assessment. Some uses are clearly instrumental: a conceptual framework can be a means for a group of assessors and users to agree on basic understandings of what features of a system are important to assess and how those features are related. Or it can be used as a one-way communication tool or device to persuade potential contributors to, and users of, an assessment to adopt a certain problem-framing favored by the proponents. Other uses fall somewhere in between, being there to inform but still open for others to question, challenge, and revise. Finally, some are really intended to be a tool for deliberation and negotiation and can be revised, fragmented, and discarded depending on the arguments made by the stakeholders involved.

3.4.1 Crossing boundaries between different forms of knowledge

Large *international* assessment efforts such as the IPCC, the MA, and the IAASTD are examples of a broad class of "boundary crossing" organizations that straddle the shifting divide between politics and science and are intended to effectively bridge boundaries between science, policy arenas, and broader society while maintaining the scientific credibility, political legitimacy, and social usefulness identified as fundamental principles in Chapter 2.

One of the practical lessons distilled from comparative studies of boundary crossing organizations concerns the production of "boundary objects," which can be maps, sketches, diagrams, historical narratives, explanatory models, and conceptual frameworks, among other things. Production of boundary objects means that scientists, policy makers, local people, civil society organizations, and other stakeholders representing different types of knowledge work together to create and use these objects; the exact form of the object is less important than the interactive process that creates it (Cash et al. 2003, Clark and Holliday 2006, Guston 2001). The conceptual framework for assessment has value as a boundary object since it can become a focal point for building effective interactions between the assessment team and stakeholders.

Studies of this boundary-crossing process have shed light on the ways "knowledge" and "action" coevolve in any multistakeholder, multidisciplinary, or multi-institutional group and how knowledge is used as the "rationale" to present actions that serve a purpose for the group. This is not easy work. A skeptical stance toward existing rationalizations is needed in order to help multiple stakeholders find productive ways out of current confusion and conflicts. Some of the existing conflicts will be based on real trade-offs and conflicting objectives; others may be at least partially linked to stereotypes, myths, and misunderstandings. For the first group of conflicts, a reorientation of objectives through negotiation and mediation probably is the best way to make progress; for the second, knowledge-based assessments can be of direct help in improving understanding.

3.4.2 A gallery of approaches

How can conceptual frameworks be made more "user-friendly" for local people? Some of the keys lie in the principles of stakeholder engagement laid out in Chapters 1 and 2 and elaborated here in section 3.2. In light of the foregoing, and based on the experiences of the sub-global assessments undertaken between 2002 and 2005 as part of the MA, it is not altogether surprising that the MA conceptual framework tended to be more readily applied at coarser than finer scales due to the difficulty of capturing local-scale interactions within the framework and the need to include stakeholder perspectives based on alternative knowledge systems and world views. Capturing these multidimensional perspectives required considerable time and the use of innovative participatory methods (MA 2005b:72–73).

The South Africa Regional Assessment (discussed in section 3.2) and the Tropical Forest Margins Assessment (discussed here) used one or more frameworks in addition to the MA framework. In a similar vein, as described in this section, the MA framework provided an important entry point for launching new assessments, such as the Northern Range and Caribbean Seas assessments. Other community-based or national assessments, such as those in Japan, Peru, and Costa Rica, have adapted the MA conceptual framework through reframing and reinterpretation in terms consistent with their communities' worldviews—often producing a radical departure from the original framework. Still others, such as Papua New Guinea (PNG), have chosen to go their own way to develop frameworks that are essentially independent from the MA version.

Tropical forest margins: Using multiple frameworks

The Partnership for the Tropical Forest Margins (known as ASB, from its former name, the Alternatives to Slash-and-Burn Program) adapted the MA conceptual framework but also continued to use its own framework for integrated natural resource management (iNRM). Figure 3.9 represents the main components of the iNRM approach and the need to ensure that work on the different components is not undertaken in isolation. The "solutions" arrived at in such an approach do not "maximize" any one element or factor in the agroecosystem; they are options for optimizing trade-offs among the different elements.

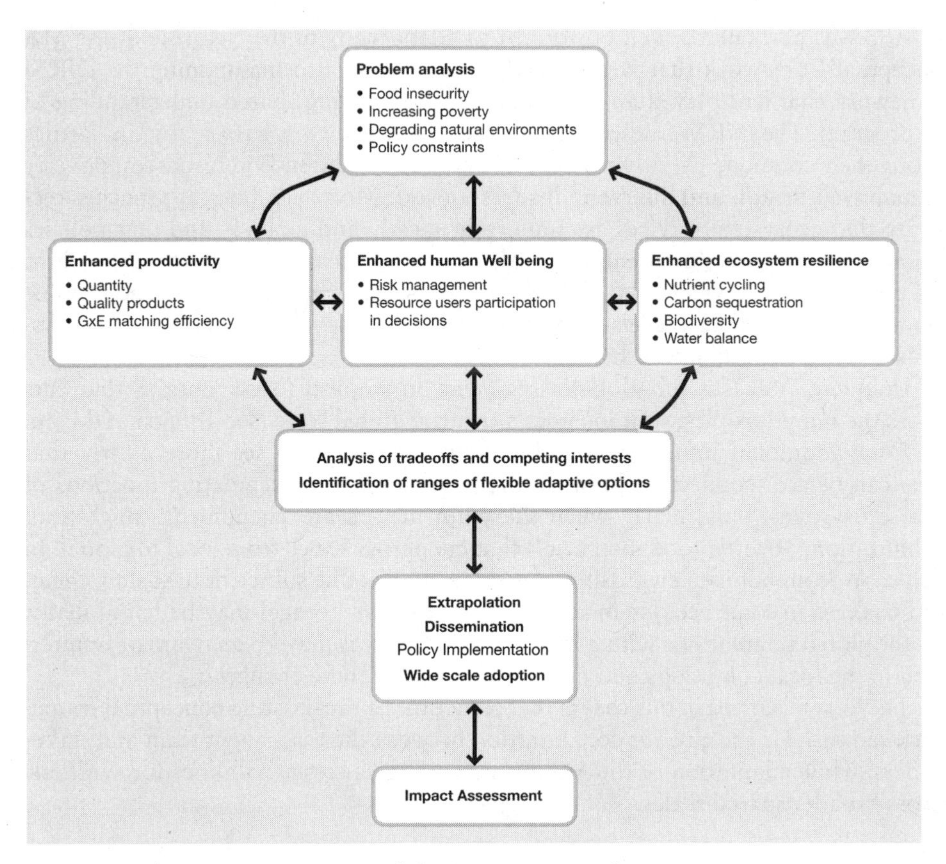

Figure 3.9. The main components of the iNRM approach.
Source: developed by Anne-Marie Izac

The concept of integrated natural resources management was developed in the context of the research undertaken by the international agricultural research centers. The kind of research conducted there evolved over time. Initially, crop improvement research (breeding "improved" germplasm to increase crop productivity) was the raison d'etre of the centers. As productivity gains occurred, scientists observed that various constraints—from inappropriate institutions and policies to limited natural resources and human resources—had to be factored in if further gains in overall production were to take place. Research on natural resources management, aimed at maintaining the properties if not the integrity of the natural resource base of agricultural systems, thus became accepted as a necessary complement to crop improvement research.

Integrated natural resources management emphasizes that the two complementary types of research need to be implemented in an integrated manner rather than simply in parallel. At the core of the iNRM approach lies the realization that there are trade-offs among different ecosystem services and among different objectives. The reductionist approach to science that is embedded in "maximizing" productivity (or profits, or water availability, or nutrient release) leads scientists to provide "solutions" that do not resolve the complex problems of agroecosystems and natural resource management because they do not address them.

ASB was particularly well positioned to adapt many of the elements of the MA conceptual framework that were directly useful while also maintaining the iNRM framework that had played an important role in building shared understanding of the program. The iNRM framework and the MA framework have many similarities among their components: driving forces, conditions and trends in resources, poverty (human well-being), and interventions (responses). Moreover, both approaches recognize that ecosystem services are important locally and globally and that policies and actions at remote (national, regional, and global) scales can affect local actions and conditions (and vice versa). But use of the MA framework enabled the ASB assessment to move to new levels by helping make more specific links between ecosystem services and human well-being.

Uniquely, ASB is a sub-global assessment on tropical forest margins that cuts across the humid tropics, so it includes a (nearly) global scale (see Tomich and Palm 2005 for additional information). This made it possible to see more clearly that there can be a disconnection between the provisioning and regulating functions of local ecosystems, particularly when the main drivers are phenomena (migration, globalization, poverty, food insecurity) that cut across scales from local to global. In such cases, community stewardship of services may not be sufficient to sustain them. And changes in some ecosystem services (e.g., carbon storage) may be felt globally, but the global populations with a stake in these services may be unaware or uninterested in the local conditions and decisions that drive these changes.

The lesson learned in this case is that sometimes a pre-existing conceptual framework can provide the glue for collaboration between the assessment team and stakeholders, while adaptation of the MA framework enhances capabilities for synthesis across broader spatial scales.

The Northern Range and Caribbean Sea: Finding the right entry points

The Northern Range and Caribbean Sea Assessments—undertaken concurrently, and jointly led by The Cropper Foundation and the University of the West Indies—generated different experiences in the application of the MA conceptual framework.

In the case of the Caribbean Sea, the authors involved in the assessment agreed that it was most meaningful to use the two priority ecosystem services provided by the sea—fisheries and tourism—as a starting point for the assessment. This decision arose from the realization that the capacity, time, and financial resources available would not allow for an assessment of all ecosystem services, driving forces, and response options required at the regional, subregional, and national levels. Further, it was agreed that because of the economic importance of fisheries and tourism to the region, the focus for formulating response options for the Caribbean Sea should be placed on these services. Having established this basis, the Caribbean Sea Assessment then proceeded to apply all components of the MA conceptual framework, using the ecosystem services box as a starting point.

In the case of the Northern Range Assessment, however, the mountain ecosystem was first subdivided into three subsystems—forest ecosystems, freshwater ecosystems, and coastal ecosystems—because it was felt that this would allow for the most effective packaging and thus assessment of the information. Because of its cross-cutting nature, biodiversity was treated separately but in the same manner as a subsystem. For each of the subsystems, the authors then identified all relevant components of the MA conceptual framework—the ecosystem services, links to human well-being, driving forces, and response options—that were applicable, and these were all assessed. Having applied this approach, the authors were more effectively able to make the linkages across the subsystems, especially related to driving forces and responses. Because of this they were able to focus a very complex and integrated set of issues in a clear, discretely laid out set of response options. When juxtaposed with the Caribbean Sea Assessment, the Northern Range Assessment used a more system-based entry point into the MA Framework.

Comparing these two experiences, it is evident that—depending on needs and issues—it is possible to apply the MA framework in different ways to answer the ultimate question: how should we respond? It is therefore extremely important that assessment practitioners gain a clear understanding of needs very early in the assessment process in order to steer the assessment methodology, including application of the conceptual framework, in the right direction.

Japan: Framing cultural concepts of ecosystems

The Japan sub-global assessment is the first of its kind in the country. It uses two terms at the core of its conceptual framework that are difficult to translate from the Japanese: *satoyama* (a traditional rural landscape) and *satoumi* (marine and coastal ecosystems with human interaction). The Japan sub-global assessment is intended to follow and apply the framework of the MA sub-global assessments by identifying the condition and trends of biodiversity, ecosystem services, and human well-being; examining the drivers of changes in these components; assessing the responses; developing conceptual models linking condition and trends, drivers, and responses; and building scenarios with a local and/or national focus on the issues of *satoyama* and *satoumi*.

While the assessment aims to provide inputs into the tenth Conference of the Parties (COP-10) of the Convention on Biological Diversity in Nagoya, Japan, in 2010, it seeks to strengthen the scientific basis for action needed to conserve and sustain *satoyama* and *satoumi* and to enhance human well-being. Although the assessment has been planned and developed since late 2006, the team in Japan spent more than a year explaining what the MA was, consulting with stakeholders regarding the planned assessment, and establishing the governance structure in a situation in which few knew about the MA or the conceptual framework it used. The development of a conceptual framework adapted to the Japan assessment still is in process, and it has taken several steps to create the platform where various types of stakeholders—users as well as experts and scientists from different disciplines—interact and share ideas.

The term *satoyama* encompasses several types of ecosystems, including secondary forests, agricultural land, ponds, grasslands, and settlements, and accounts for approximately 40% of the land area of Japan. *Satoyama* also is a way of life and a perspective on interactions between humans and the environment. More recently, this concept of interaction in *satoyama* has been expanded to include marine and coastal ecosystems, giving rise to the idea of *satoumi*. Although there still is some debate about this, the mosaic patterns of ecosystems appear to maintain a high level of biodiversity. While the government of Japan has become committed to promoting the term *satoyama* to the international community, especially at COP-10, there is no common definition of *satoyama* and there also are various similar phrases and terminologies in Japanese. This is because numerous groups and individuals have attempted to define *satoyama* from their own background and interests, and some refer to it as secondary forests.

The assessment is trying to look into various definitions that include different types of ecosystems and different area categories as well as historical narrative contexts of *satoyama* and *satoumi* to identify the unit and the scale of mosaic-type systems with human interaction and also to determine the linkages to nature and to urban areas, given that *satoyama* usually exist between primary forests and cities. In addition, one of the major issues concerning *satoyama* and *satoumi* in Japan is their underuse (versus an overuse of many ecosystems or their services) through aging and the loss of population in many localities. This is expected to provide some views on pressure on ecosystems that would be different from most assessments in developing countries—in particular in terms of pressures on mountain and forest ecosystems—and to suggest the extent to which humans should intervene to maintain the ecosystems that have been abandoned and, as a result, are undergoing natural succession.

Although these are distinctively Japanese terms and they are difficult to translate in ways that do justice to their full meaning, these types of multi-ecosystem landscapes are not unique to Japan. Such complex landscape mosaics are found throughout the world. Top-down global approaches consistently miss these important integrative elements at the landscape scale. The Japanese sub-global assessment may produce insights that could be useful to other assessment teams.

Peru: Developing locally relevant conceptual frameworks

Indigenous communities conducting an assessment in the Vilcanota region of Peru created a conceptual framework using the Quechua understanding of ecological and social relationships (see Figure 3.8 in section 3.3.6). Within the Quechua vision of the cosmos, concepts such as reciprocity (*Ayni*), the inseparability of space and time, and the cyclical nature of all processes (*Pachakuti*) are important components of the Inca definition of ecosystems. Love (*Munay*) and working (*Llankay*) bring humans to a higher state of knowledge (*Yachay*) about their surroundings and are therefore key concepts linking Quechua communities to the natural world. *Ayllu* represents the governing institutions that regulate interactions between all living beings. These concepts have existed since the time of the Incas and are deeply ingrained within Quechuan communities.

The communities, and particularly those leading the assessment, felt that the assessment work would have less relevance if the MA conceptual framework were used as a base for the assessment process. The framework that was developed and used has similarities with the MA's, but the divergent features were considered important to the Quechua people conducting the assessment. The various similarities and differences between the conceptual frameworks were discussed to build understanding of the broader spectrum of perceptions about the relationship between ecosystem services, human well-being, and drivers of change.

The Vilcanota conceptual framework also includes multiple scales (*Kaypacha*, *Hananpacha*, *Ukupacha*); however, these represent both spatial scales and the cyclical relationship between the past, present, and future. Inherent in this concept of space and time is the adaptive capacity of the Quechua people, who welcome change and have developed resilience through an adaptive learning process. (It is recognized that current rates of change may prove challenging to the adaptive capacities of the communities.) The cross shape of the Vilcanota framework diagram represents the *Chakana*, the most recognized and sacred shape to Quechua people, and orders the world through deliberative and collective decision making that emphasizes reciprocity. *Pachamama* is similar to a combination of the "ecosystem goods and services" and "human well-being" components of the MA framework. *Pachakuti* is similar to the MA drivers (both direct and indirect). *Ayllu* (and *Munay*, *Yachay*, and *Llankay*) may be seen as responses and are more organically integrated into the cyclic process of change and adaptation.

In the Vilcanota assessment, the Quechuan communities directed their work process to assess the conditions and trends of certain aspects of the *Pachamama* (focusing on water, soil, and agrobiodiversity), how availabilities of these goods and services are changing, the reasons behind the changes, the effects on the other elements of the *Pachamama*, how the communities have adapted and are adapting to the changes, and the state of resilience of the Quechua principles and institutions for dealing with these changes in the future.

Developing the local conceptual framework from a base of local concepts and principles, as opposed to simply translating the MA framework into local terms, allowed local communities to take ownership of the assessment process and gave them the power both to assess the local environment using their own knowledge and principles of well-being and to seek responses to problems within their own cultural and spiritual institutions. Strong involvement from the Quechua community in the process produced a framework that reflects local views on the place of humans within the natural world but that also has many similarities with the MA framework. The similarities were important to discuss in order to build a process that matched the goals of the MA.

Costa Rica: Adaptation of local ecological knowledge for resource management

The Asociacion Ixacavaa de Desarrollo y Informacion Indigena began the Bajo Chirripo assessment with the idea of developing a management plan for the community's resources. Through discussions and meetings with community members, it soon became apparent that in the past a strict "management plan" had existed and was based on norms and beliefs regarding interactions between humans and their environment. The concept of reciprocity between humans and the rest of the environment was key. The erosion of traditional knowledge and of the norms and beliefs that have guided how the Cabecar have traditionally managed their resources became the major impetus for developing a local, participatory assessment process.

Ixacavaa therefore focused their ecosystem assessment process on the recovery of lost knowledge that in the past safeguarded the integrity of the environment and ensured the sustainability of human activities. Because of this, it was important to make sure that the conceptual framework used was one that fit this goal and emphasized the importance of local knowledge. The discussion and development of a conceptual framework was also an opportunity to initiate discussions within the community about lost knowledge and how to recuperate it.

In Bajo Chirripo, the basic concepts of the MA conceptual framework were understood to some extent, but it was nevertheless quite foreign to community members' way of thinking. Recognizing a need to understand these concepts in the language and from the perspective of the community, the assessment team and the Cabecar community invested considerable time in revising the conceptual framework. A full local framework that could be used in assessment was not completed, but components were identified and discussed in the community.

One of the components identified as central to ecosystem management is the Cabecar conical shaped house, which represents the natural world. The flip side (also a conical house) is the spirit world and is equally important. Communities had noticed a big decline in the number of animals and the quantity of important natural resources available to them in their territory. They explain this by saying that the animals have left the natural world and are hiding in the spirit world. When humans begin to act more responsibly and with greater reciprocity, it is believed that the animals will return to the natural world. Because of the discussion of these ideas in the process of the assessment, community members in one village decided to build a conical house (which can also be physically constructed on earth and becomes a spiritual icon to the communities), which has now been completed.

Papua New Guinea: An independent approach

Papua New Guinea is a nation of indigenous peoples, with several hundred languages and cultures, who generally like to think of themselves as "customary landowners." Members of this society are not very keen on flow charts as a way to think about environmental issues, and although there are many different forms of traditional environmental knowledge, they do not seem to constitute "systems" of the sort portrayed in Figure 3.8. In this context, it was felt that the best way to adapt the MA conceptual framework was to use the simplest of all diagrammatic forms, the four-cell matrix, as a way to explore the difference between scientific and indigenous perspectives on the definition of crucial terms in the environmental policy process.

The first home-grown diagram in the conceptual framework chapter of the PNG assessment therefore offers four ways of defining an ecosystem (see Figure 3.10). It recognizes the ambiguity already present in the scientific perspective of the MA conceptual framework and then suggests a parallel ambiguity in the "political" perspective that would make sense to a nation of customary landowners who have good reason to consider the territorial domain of a traditional political community as an ecosystem in its own right, one that is internally divided by landscape elements that are normally distributed in equal measure between smaller groups within the community.

	Geographical Perspective	Biological Perspective
Scientific Perspective	Physical Environments	Biological Communities
Political Perspective	Territorial Domains	Landscape Elements

Figure 3.10. Defining an ecosystem.
Source: PNG sub-global assessment (unpublished)

The same logic is later applied to the identification of significant species within an ecosystem in order to get a handle on the evident gap between the way in which scientists and local people think about biodiversity or "life on Earth" (see Figure 3.11). Biologists distinguish between an endemic species, which is unique to a certain type of ecosystem and whose survival therefore depends on the survival of that ecosystem, and a keystone species, which makes a unique contribution to the survival of a certain type of ecosystem, even if it is not endemic (Mills et al. 1993). A keynote species, by contrast, is one whose services are essential to the survival of a specific form of traditional or indigenous culture, while a totemic species (at least in the Melanesian context) is one whose services are recognized in magic and mythology and hence in the value that local people attribute to its reproduction.

	Essential	Important
Scientific Perspective	Keystone Species	Endemic Species
Indigenous Perspective	Keynote Species	Totemic Species

Figure 3.11. Identifying significant species within an ecosystem.
Source: PNG sub-global assessment (unpublished)

Another four-cell matrix is applied to the analysis of values, drivers, and responses (see Figure 3.12), establishing a contrast between the production systems and management regimes that operate at a community scale (where local people define ecosystems as territorial domains) and those that operate at a wider social scale.

	Community	Society
Production Systems	Indigenous Systems	Industrial Systems
Management Regimes	Local Regimes	Sectoral Regimes

Figure 3.12. Analyzing values, drivers and responses.
Source: PNG sub-global assessment (unpublished)

This way of thinking about the world has obvious drawbacks if it leads people to put things into one of four pigeonholes to which they do not really belong. But that is not the point. In the last case the link made between indigenous production systems and traditional territorial domains allows a number of useful insights. First, it does away with the less helpful distinction often made between the cash and subsistence sectors of a national economy in the valuation of ecosystem services, for many of the products of indigenous production systems are actually sold in domestic and international markets. Second, indigenous knowledge can be thought of as a set of loosely connected practical understandings embedded in different forms of production rather than a picture of the world as a whole, and this does seem to make more sense of the way that Melanesians (and many other people) deal with their natural environment. Third, the innovative and adaptive nature of what might otherwise be called "traditional" production systems is clearly recognizable, avoiding the romantic and misguided idea that customary landowners simply follow in the footsteps of their ancestors. Although it is then possible to ask how indigenous and industrial production systems may compete or collaborate in specific economic sectors (forestry or agriculture, for example), this still allows recognition of the existence of production systems (like dive tourism or illegal fuelwood harvesting) that occupy an informal or intermediate space between these two categories.

The PNG assessment is a reminder that conceptual frameworks, knowledge systems, and computer programs are all built out of basic binary distinctions. Yet there is no reason to assume that distinctions that make sense in one context or to one group of people can be absorbed into a conceptual framework that makes sense in all contexts and to all possible users of an ecosystem assessment. The best way to test the validity and usefulness of any conceptual framework is to check the significance of such distinctions to the people who "relate to nature" or "use ecosystem services" in the place where an assessment is conducted.

References

Arrow, K. J., P. Dasgupta, and K.-G. Maler. 2004. Evaluating projects and assessing sustainable development in imperfect economies. In *The economics of non-convex ecosystems.*, ed. P. Dasgupta and K.-G. Maler. Dordrecht, the Netherlands: Kluwer Academic Publishers.

Baumol, W. J., and W. E. Oates. 1988. *The theory of environmental policy*. Cambridge, U.K.: Cambridge University Press.

Bell, S., and S. Morse. 1999. *Sustainability indicators: Measuring the immeasurable*. London: Earthscan Publications.

Bell, S., and S. Morse. 2003. *Measuring sustainability: learning by doing*. London: Earthscan Publications.

Berger, A. R., and R. A. Hodge. 1998. Natural change in the environment: A challenge to the pressure-state-response concept. *Social Indicators Research* 44 (2): 255–65.

Carney D. 1998. *Sustainable rural livelihoods: what contribution can we make*? DfID, London

Cash, D.W. W.C. Clark, F. Alcock, N.M. Dickson, N. Eckley, D. Guston, J. Jäger, and R. Mitchell. 2003. Knowledge systems for sustainable development. *Proceedings of the National Academy of Sciences of the United States of America* 100: 8086–8091.

Chapin, F. S. III, E. S. Zavaleta, V. T. Eviner, R. L. Naylor, P. M. Vitousek, H. L. Reynolds, D. U. Hoope et al. 2000. Consequences of changing biodiversity. *Nature* 405:234–42.

Clark, W.C., and L. Holliday. 2003. Linking Knowledge with Action for Sustainable Development. National Research Council of the National Academies: Washington, D.C.

Dore, J. 2007. Multi-stakeholder platforms (MSPS): unfulfilled potential. In *Democratizing water governance in the Mekong region*, ed. L. Lebel, J. Dore, R. Daniel, and Y. Koma, 197–226. Chiang Mai, Thailand: Mekong Press.

Folke, C., S. Carpenter, T. Elmquist, L. Gunderson, C. S. Holling, and B. Walker. 2002. Resilience and sustainable development: Building adaptive capacity in a world of transformations. *Ambio* 31 (5): 437–40.

Geist, H. J. 2005. *The causes and progression of desertification*. Ashgate Studies in Environmental Policy and Practice. Aldershot, U.K.: Ashgate.

Geist, H. J. 2006. Change or collapse? A theoretical approach to global environmental change and land use in rainforest and arid zone hotspots. (In German.) *Geographische Zeitschrift* 94 (3):143–59.

Geist, H. J. and E. F. Lambin. 2002. Proximate causes and underlying driving forces of tropical deforestation. *BioScience* 52 (2): 143–49.

Geist, H. J. and E. F. Lambin. 2004. Dynamic causal patterns of desertification. *BioScience* 54 (9): 817–29.

Geist, H. J., E. Lambin, C. Palm, and T. Tomich. 2006. Agricultural transitions at dryland and tropical forest margins: Actors, scales and trade-offs. In *Agriculture and climate beyond 2015: A new perspective on future land use patterns*, ed. F. Brouwer and B. A. McCarl, 53–73. Dordrecht, Netherlands: Springer-Verlag.

Gunderson, L. H., and C. S. Holling. 2002. *Panarchy: Understanding transformations in human and natural systems*. Washington, DC: Island Press.

Guston, D. 2001. Boundary organizations in environmental policy and science: an introduction. *Science, Technology and Human Values* 26: 399–408.

Heinz Center. 2006. *Filling the gaps: priority data needs and key management challenges for national reporting on ecosystem condition*. Washington, DC: Heinz Center for Science, Economics and the Environment.

IPCC (Intergovernmental Panel on Climate Change). 2007. *Climate change 2007—IPCC fourth assessment report*. Cambridge, U.K.: Cambridge University Press.

Jerneck, A., and L. Olsson. 2007. More than Trees: Contextualising agro-forestry and identifying opportunities in subsistence farming. Internal report to World Agroforestry Centre, December.

Kasemir, B., M. B. A. Asselt, and G. Durrenberger. 1999. Integrated assessment of sustainable development: Multiple perspectives in interaction. *International Journal of Environment and Pollution* 11 (4): 407–25.

Kuhn, T. 1962. *The structure of scientific revolutions*. Chicago: University of Chicago Press.

Lambin, E. F., H. J. Geist, and E. Lepers. 2003. Dynamics of land use and cover change in tropical and subtropical regions. *Annual Review of Environment and Resources* 28:205–41.

Lambin, E. F., H. J. Geist, and R. R. Rindfuss. 2006. Introduction: Local processes with global impacts. In *Land-use and land-cover change: Local processes and global impacts*. Global Change—The IGBP Series, ed. E. F. Lambin and H. J. Geist, 1–8. Berlin: Springer-Verlag.

Lambin, E. F., H. J. Geist, J. F. Reynolds, and D. M. Stafford Smith. 2007. Integrated human–environment approaches of land degradation in drylands. In *Sustainability or collapse? An integrated history and future of people on earth*. Dahlem Workshop Series 96, ed. R, Costanza, L. J. Graumlich, and W. Steffen, 331–39. Cambridge, MA: The MIT Press.

Lebel, L. 2006. The politics of scale in environmental assessment. In *Bridging Scales and Knowledge Systems: Concepts and Applications in Ecosystem Assessment*, ed. W. V. Reid, F. Berkes, T. J. Wilbanks, and D. Capistrano, 37–57. Washington, DC: Island Press.

Lebel, L., and P. Garden. 2008. Deliberation, negotiation and scale in the governance of water resources in the Mekong region. In *Adaptive and integrated water management: coping with complexity and uncertainty*, ed. C. Pahl-Wostl, P. Kabat, and J. Möltgen, 205–25. Berlin: Springer.

Lebel, L., R. Daniel, N. Badenoch, and P. Garden. 2008. A multi-level perspective on conserving with communities: Experiences from upper tributary watersheds in montane mainland southeast Asia. *International Journal of the Commons* 1:127–54.

Lee, K. N. 1999. Appraising adaptive management. *Conservation Ecology* 3 (2): 3. [online] URL: http://www.ecologyandsociety.org/vol3/iss2/art3.

Ludwig, D. B., B. Walker, and C. S. Holling. 1997. Sustainability, stability, and resilience. *Conservation Ecology* 1 (1): 7.

Machlin, G. E., and J. E. McKendry. 2005. *The human ecosystem as an organizing concept in ecosystem restoration*. HESG Publications 05-01. Moscow, ID: University of Idaho.

Meter, K. 1999. *Neighborhood sustainability indicators guidebook*. Minneapolis, MN: Crossroads Resource Center.

Michon, G., H. de Foresta, P. Levang, and F. Verdeaux. 2007. Domestic forests: a new paradigm for integrating local communities' forestry into tropical forest science. *Ecology and Society* 12(2): 1. [online] URL: http://www.ecologyandsociety.org/vol12/iss2/art1/.

MA (Millennium Ecosystem Assessment). 2003. *Ecosystems and human well-being: A framework for assessment*. Washington, DC: Island Press.

MA. 2005a. *Ecosystems and human well-being, Vol. 1: Current state and trends*. Washington, DC: Island Press.

MA. 2005b. *Ecosystems and human well-being, Vol. 4: Multiscale assessments*. Washington, DC: Island Press.

Mills, L. S., M. E. Soule, and D. F. Doak. 1993. The keystone-species concept in ecology and conservation. *BioScience* 43 (4): 219–24.

Nelson, G. C., E. Bennett, A. A. Berhe, K. Cassman, R. DeFries, T. Dietz, A. Dobermann et al. 2006. Anthropogenic drivers of ecosystem change: An overview. *Ecology and Society* 11 (2): online journal.

Ness, B., S. Anderberg, and L. Olsson. 2008. Structuring problems of unsustainability in the Swedish sugar sector with the multilevel DPSIR framework. *Sustainability Science*. Accepted for publication.

Newell, B. C., C. L. Crumley, N. Hassan, E. F. Lambin, C. Pahl-Wostl, A. Underdal, and R. Wasson. 2005. A conceptual template for human–environment research. *Global Environmental Change* 15 (4): 299–307.

Pirrone, N., G. Trombino, S.Cinnirella, A. Algieri, G. Bendoricchio, and L. Palmeri. 2005. The Driver-Pressure-State-Impact-Response (DPSIR) approach for integrated catchment–coastal zone management: preliminary application to the Po catchment, Adriatic Sea coastal zone system. *Regional Environmental Change* 5: 111–137.

Reid, R. S., T. P. Tomich, J. C. Xu, and A. S. Mather. 2006. Linking land-change science and policy: Current lessons and future integration. In *Land-use and land-cover change: Local processes and global impacts*, ed. E. F. Lambin and H. J. Geist, 157–71. Berlin: Springer.

Reid, W., et al. 2006. *Bridging Scales and Knowledge Systems: Linking Global Science and Local Knowledge in Assessments*. Washington, D.C.: Island Press.

Resilience Alliance and Sante Fe Institute. 2008. *Thresholds and regime shifts in ecological and social-ecological systems*. Resilience Alliance.

Reynolds, J. F., and D. M. Stafford Smith, eds. 2002. *Global desertification: Do humans cause deserts*. Dahlem Workshop Series 88. Berlin: Dahlem Unversity Press.

Reynolds, J. F., D. M. Stafford Smith, E. F. Lambin, B. L. Turner II, M. Mortimore, S. P. J. Batterbury, T. E. Downing et al. 2007. Global desertification: Building a science for dryland development. *Science* 316 (5826): 847–51.

Schellnhuber, H.-J., M. K. B. Lüdeke, and G. Petschel-Held. 2002. The syndromes approach to scaling/describing global change on an intermediate functional scale. *Integrated Assessment* 3 (2/3): 201–19.

Senge, P., R. Ross, C. Roberts, B. Smith, and A. Kleiner. 1994. *The fifth discipline fieldbook: Strategies and tools for building a learning organization*. London: Nicholas Brealy.

Simon, H. A. 1962. The architecture of complexity. *Proceedings of the American Philosophical Society* 106 (6): 467–82.

Smeets E., R. Weterings. 1999. Environmental indicators: typology and overview. Technical Report 25, European. Environmental Agency, Copenhagen, Denmark. http://reports.eea.eu.int:80/TEC25/en/tech25text.pdf.

Svarstad, H., L. Kjerulf Petersen, D. Rothman, H. Siepel, and F. Watzold. 2008. Discursive biases of the environmental research framework DPSIR. *Land Use Policy* 25: 116–125.

Spreng, D., and A. Wils. 1996. Indicators of sustainability: Indicators in various scientific disciplines. *Multidimensional approaches to sustainability: The framing project*. Alliance for Global Sustainability: 1–27.

Stern, D. 1995. *The capital theory approach to sustainability: A critical appraisal*. Boston University Center for Energy and Environmental Studies Working Papers Series No. 9501: 1–44.

Tomich, T. P., and C. A. Palm et al. 2005. *Forest and agroecosystem tradeoffs in the humid tropics: A crosscutting assessment by the Alternatives to Slash-and-Burn consortium conducted as a sub-global component of the Millennium Ecosystem Assessment*. Nairobi: Alternatives to Slash-and-Burn Programme.

Tomich, T. P., D. Timmer, J. C. Alegre, V. Areskoug, D. Cash, A. Cattaneo, P. Ericksen et al. 2007. Integrative science in practice: Process perspectives from ASB, the Partnership for the Tropical Forest Margins. *Agriculture Ecosystems and Environment* 121 (3): 269–86.

UNEP (U.N. Environment Programme). 2004. Synthesis of responses on strengthening the scientific base of the United Nations Environment Programme, Report by the Executive Director. Nairobi.

UNEP. 2007. *Global Environment Outlook–4*. Nairobi: UNEP.

van Noordwijk, M., D.A. Suyamto, B. Lusiana, A. Ekadinata, and K. Hairiah. 2008. Facilitating agroforestation of landscapes for sustainable benefits: tradeoffs between carbon stocks and local development benefits in Indonesia according to the FALLOW model. *Agriculture Ecosystems and Environment* 126: 98–112.

Victor, P. A. 1991. Indicators of sustainable development: Some lessons from capital theory. *Ecological Economics* 4:191–213.

Walters, C., and C. S. Holling. 1990. Large-scale management experiments and learning by doing. *Ecology* 71 (6): 2060–68.

Waltner-Toews, D., and J. Kay. 2005. The evolution of an ecosystem approach: The diamond schematic and an adaptive methodology for ecosystem sustainability and health. *Ecology and Society* 10 (1): 38.

Warner, J. F. 2006. More sustainable participation? Multi-stakeholder platforms for integrated catchment management. *Water Resources Development* 22:15–35.

WCED (World Commission on Environment and Development). 1987. *Our Common Future*. Oxford: Oxford University Press.

4

Assessing State and Trends in Ecosystem Services and Human Well-being

Robert Scholes, Reinette Biggs, Cheryl Palm, and Anantha Duraiappah

What is this chapter about?

This chapter provides practical guidance on gathering, evaluating, and presenting information related to the supply and consumption of ecosystem services and the status of human well-being, especially to the extent that it depends on ecosystem services. The chapter suggests the types of measurements that can be used as indicators of both ecosystem services and human well-being. It shows how information from a variety of sources can be integrated to provide coherent and robust insights into trends in the adequacy and security of ecosystem service provision. It suggests ways of communicating the information in an effective and responsible manner, including where information is incomplete or uncertain.

4.1 How to set the scope of a condition and trend assessment

Section's take-home messages

- It is assumed that the geographical scope and content of an assessment have already been broadly determined through the stakeholder engagement processes.
- One part of the scope-setting exercise that is particular to the condition and trend assessment is agreeing on the time period for assessing conditions and for analyzing trends. It is helpful to think of the period for assessing trends as "the relevant past to the predictable future."

The "state" of human well-being or of ecosystem services is a snapshot of its condition in a given area and at a particular time, usually the present or recent past. State is not synonymous with "health," "integrity," or "degradation" of ecosystems, though all of those related concepts may be part of an ecosystem condition assessment. State can be measured in many ways, including, for instance, the yield of a service, the stock of natural capital that permits that yield, the economic value of either, and various indicators of human well-being. "Trend" is an analysis of the change in state over time.

It is assumed that the geographical scope and content of an assessment have already been broadly determined through the stakeholder engagement processes described in detail in Chapter 2. An awareness of the technical issues and data implications for condition and trend analysis is a necessary input to the discussion that leads to this decision, which implies that setting the scope is best done iteratively. Either the technical experts or the stakeholders can make the first proposal, which then needs to be debated and modified successively by both users and producers of the assessment until a workable consensus is reached.

A part of the scope-setting exercise that is particular to the condition and trend assessment is agreeing on the time period for assessing conditions and for analyzing trends. A condition assessment applies to a particular time, which is seldom an exact instant but is a relatively short period. Many statistics are collected on an annual basis, but either there is considerable interannual variation or the trends are quite slow, making reporting on longer periods (such as a decade) more useful. Unlike the objective of trend analysis, the goal is to detect change between periods, not within them. The window must be long enough to collect reliable data, but not so long that too much change occurs within it.

It is helpful to think of the period for assessing trends as "the relevant past to the predictable future." Ecosystem services typically exhibit some "path dependency"; in other words, their current levels depend to some degree on their history. But it is not necessary to push the record as far back as is technically possible unless the path dependency of the particular services and processes being examined requires it. A single human generation (three decades) is often enough. It is also untrue that extrapolation into the future is impossible. Ecosystem service and human well-being trends are generally quite predictable for a limited period, and all rational managers make such extrapolations. The length of that period varies, depending on the processes involved—"slow processes" such as the growth of long-lived trees can be extrapolated for decades, whereas "fast processes" such as deforestation may only be predictable a few years ahead. For times beyond these limits, fundamental uncertainties make such predictions indeterminate, and scenario analysis is the appropriate tool (see Chapter 5).

4.2 How to select what to measure

Section's take-home messages

- An indicator, which is usually quantitative, is a single variable with some logical connection to the process or object of concern. Compound indices are measures made up of several different indicators, combined in a particular way to increase their sensitivity, reliability, or ease of communication.
- Assessments of ecosystem condition rely on sets of indicators. Ideally, individual indicators should be policy relevant, scientifically sound, simple to calculate and easy to understand, practical and affordable, sensitive to relevant changes, suitable for aggregation and disaggregation, and usable for projections of future scenarios.
- The different sources of information for the assessment include peer-reviewed literature, statistical databases, maps and remotely sensed images, computer models, indigenous technical and traditional knowledge, and Internet sources.
- Some things do not fit easily into an ecosystem service framework. Biodiversity, for

example, is generally not an ecosystem service itself but a necessary condition for such services to be delivered. There are numerous ways to measure and express biodiversity.
- Since ecosystem services are place and time specific and do not necessarily aggregate upward in a simple additive way or disaggregate downward by simple proportionality, care must be taken in relation to scale questions in both time and space. It is important to work at an appropriate scale for both ecological and human processes.

4.2.1 Metrics, indicators, and indices

A metric is a quantitative measurement. An indicator can be anything that contains useful information (a bad smell is an indicator of rotten food, for instance), but in the assessment context it is usually quantitative. Indicators are single variables with some logical connection to the process or object of concern—they reflect in some unambiguous way its status, causes (drivers), or outcome. It is quite common (but not essential) to "index" such variables to a particular reference state, for example by

- Setting them to a value of 100 at a particular time, and then expressing all previous and subsequent values relative to that number;
- Rescaling them between 0 and 1 by subtracting the minimum value from the observed value and dividing by the difference between the minimum and maximum value (a variant of this is expressing the observed value as a percentage of the maximum possible value); or
- Normalizing them by subtracting the value from the mean, and dividing by the standard deviation.

When indexed in this way, the indicators are nondimensional and can be added together, for instance in compound indexes. Compound indexes are measures made up of several different indicators, combined in a particular way to increase their sensitivity, reliability, or ease of communication. For instance, a geometric mean is often used rather than a simple arithmetic mean, to prevent very high values in one component from dominating the calculation.

Compound indexes should be used with great care: they run the risk of canceling out underlying trends when the component indicators change in different directions, and they always have value judgments encoded in their weighting rules, either explicitly or implicitly. Even "unweighted" indexes suffer this problem, since they assume that all factors are equally important.

The attributes of good indicators are given in Box 4.1. There are advantages to adopting indicators that have already been agreed to by a political and technical process. They can be adopted by themselves or in conjunction with other indicators as appropriate. For example, a lengthy process of discussion and experimentation has resulted in agreed lists of key indicators for monitoring biodiversity under the Convention on Biological Diversity (a summary is given in Table 4.1). Similarly, for human well-being a framework of eight goals, 18 targets, and 48 indicators to measure progress toward the Millennium Development Goals was adopted by a consensus of experts from the United Nations Secretariat and the International Monetary Fund (IMF), the Organisation for Economic Co-operation and Development (OECD), and the World Bank (see section 4.4).

Box 4.1. Principles for choosing indicators

Individual indicators should ideally be:

1. Policy relevant
Indicators should provide policy-relevant information at a level appropriate for decision making. Where possible, indicators should allow for assessment of changes in ecosystem status related to baselines and agreed policy targets.

2. Scientifically sound
Indicators should be based on clearly defined, verifiable, and scientifically acceptable data, collected using standard methods with known accuracy and precision or based on traditional knowledge that has been validated in an appropriate way.

3. Simple to calculate and easy to understand
Indicators should provide clear, unambiguous information that is easily understood. It is important to jointly involve policy makers, major stakeholders, and experts in selecting or developing indicators to ensure that the indicators are appropriate and widely accepted.

4. Practical and affordable
Obtaining data on the indicator should be practical and affordable.

5. Sensitive to relevant changes
Indicators should be sensitive and able to detect changes at time frames and spatial scales that are relevant to the decision making. At the same time, they should be robust to measurement errors or random environmental variability in order to prevent "false alarms." The most useful indicators are those that can detect change before it is too late to correct the problems.

6. Suitable for aggregation and disaggregation
Indicators should be designed in a manner that facilitates aggregation to higher scales or disaggregation to lower scales in space or time, for different users. Indicators that can be expressed in relation to ecosystem boundaries as well as political boundaries are very useful.

7. Usable for projections of future scenarios
Indicators that allow cause-effect relationships to be quantified and projected forward allow for scenario analyses. This can enable evaluation of alternative policy options or management strategies.

Sets of indicators: No single indicator can provide information on all policy-relevant changes. Assessments of ecosystem condition therefore rely on sets of indicators. Ideally, the chosen set should include a relatively small number of individual indicators and be representative of the relevant issue. The smaller the total number of indicators, the lower the cost and the easier it is to communicate the findings to policy makers and the public. However, the set of indicators should not be so small or simple that they ignore important aspects of the issue being assessed.

Source: Based on CBD 2003.

Table 4.1. Indicators developed for the Convention on Biological Diversity, in relation to their 2010 target

Focal Area	Indicator
Status and trends of the components of biological diversity	• Trends in extent of selected biomes, ecosystems, and habitats • Trends in abundance and distribution of selected species • Coverage of protected areas • Change in status of threatened species • Trends in genetic diversity of domesticated animals, cultivated plants, and fish species of major socioeconomic importance
Sustainable use	• Area of forest, agricultural, and aquaculture ecosystems under sustainable management • Proportion of products derived from sustainable sources • Ecological footprint and related concepts
Threats to biodiversity	• Nitrogen deposition • Trends in invasive alien species
Ecosystem integrity and ecosystem goods and services	• Marine Trophic Index • Water quality of freshwater ecosystems • Trophic integrity of other ecosystems • Connectivity / fragmentation of ecosystems • Incidence of human-induced ecosystem failure • Health and well-being of communities who depend directly on local ecosystem goods and services • Biodiversity for food and medicine
Status of traditional knowledge, innovations and practices	• Status and trends of linguistic diversity and numbers of speakers of indigenous languages • Other indicator of the status of indigenous and traditional knowledge
Status of access and benefit-sharing	• Indicator of access and benefit-sharing
Status of resource transfers	• Official development assistance provided in support of the Convention • Indicator of technology transfer

Source: UNEP/CBD/COP/7/21/Part 2, Decision VII/30; summarized in Balmford et al. (2005). Available at www.biodiv.org/decisions

4.2.2 Different sources of information

The purpose of assessments is to collate, organize, analyze, evaluate, and present information. The sources of this information can be extremely diverse, but they should all pass the same basic rules of admissibility:

- *Can they be traced?* If nobody can get access to the fundamental information other than those who prepared it, it has little credibility. This means that the

sources have to be attributed, in the public domain (i.e., not secret, although some restrictions on making the information freely available can be applied in rare cases), and accessible (in a library or archive somewhere).
- *Can they be tested?* This is a hallmark of the "scientific method." Ecosystem assessments are part of the scientific knowledge system, even when they draw on nonformal technical knowledge. If the information is of such a nature that it could never be validated or it has no generality or predictive capacity (for instance, if every case is completely unique), it does not lend itself to assessment.

Peer-reviewed literature

Peer-reviewed literature is the main source of assessment information, precisely because it represents information that is already in the public domain and has already been tested. Assessments are not meant to generate new primary knowledge (i.e., assessments are not research projects, although they often spawn such projects). They are meant to draw mostly on existing information. Passing the test of peer review is not a guarantee of correctness, but it does increase the confidence with which material can be used. "Peer-reviewed" usually means published in an academic journal or book, but other pathways are possible. Any credible and documented process by which the information has undergone reasonably independent checking is acceptable.

Correct attribution ("referencing") of source material is essential for traceability. Using a consistent format will ensure that all the necessary information is present (see Box 4.2). It is essential to reflect the information used in an honest, fair, and balanced way. In other words, it is not appropriate to quote sources that have not been read or to give selective quotations that misrepresent the overall message of the source. It is the job of the review editors to be vigilant about such abuses. Correct attribution of information and ideas is an issue of professional ethics—it is not acceptable to simply cite a recent source when the original idea preceded it. The reference lists should contain all author names (not just et al.). And cultural, language, discipline, and geographic bias should be minimized in any literature consulted.

Often assessments are greatly assisted by being able to use "gray literature." This is material in some form of document, such as an internal report or a low-circulation periodical, that is in principle already in the public domain but is effectively unavailable to the broader community. Such material may be used, provided a copy is placed in a public domain repository, such as library or an archive.

Assessments wish to use the most up-to-date information, but some of this may be in the form of "submitted," "in review," or "in press" papers. In press articles (which have been accepted by the peer review process and are simply awaiting publication) may be cited. Submitted or in review papers can be included in drafts of the assessment, but if they are not published or in press by the time the assessment is in final proofs, they must be removed and any conclusions critically dependent on them must be revised.

Statistical databases

Often the quantitative information in assessments is based on databases (usually electronic, but sometimes on paper) that are not peer-reviewed or published in the same sense as the literature just described. The correct acknowledging citation of the

Box 4.2. Referencing information sources

The ability to trace information to its source is a key indicator of credibility. What follows is not a prescribed reference style, but a list of essential information for different types of sources. Use whatever reference style is most appropriate for the assessment's intended audience. For instance, some users prefer extensive footnotes or numbered endnotes to the "scientific" style of embedding author and date in the text.

1. **An article in a periodical**: author(s), year, title of paper, name of journal, volume, and page range.
2. **A book or report**: authors(s), year, title of volume, ISBN, publisher, publisher's address, and number of pages.
3. **A chapter in a book**: author(s), year, title of chapter, editors of book, title of book, publisher, publisher's address, and page range.
4. **A conference proceeding**: author(s), year, title of presentation, name of conference or proceeding, date and place, organizing institution, address, and page range in proceeding or published abstract.
5. **A personal communication**: (i.e., something told by an informant). Name and initials of the person, date, and an address or email contact. Note: this is generally a low-credibility source and should be used sparingly unless there is no alternative. If the person has credentials or experience that give them credibility, note that.
6. **A database**: author(s), date of access, version, name of database, and responsible institution.
7. **A Web site**: html address, date of access, and author (where apparent).
8. **A photograph**: photographer, date, and place (for repeat photography, also the lens used and direction of the photo).

databases (i.e., traceable and fair to the people who have put the effort into them) is important (see Box 4.2 for an example). Note that although the copyright of the data resides with the assessment, if the data have been altered substantially (expressed in a different form, for instance), it is still necessary for reasons of courtesy and transparency to cite the originator. Since many such databases are "live," it is important to record the date on which the information was retrieved.

Determining the quality of data in databases is important. One approach is to use information from different sources and see if it is in reasonable agreement. A second check is for internal consistency. Do the row and column totals add up correctly? Are the location data in fact in the place where the study is meant to cover? Be aware of (and acknowledge) potential biases, uncertainties, and gaps in the data, especially where independent corroboration is difficult. Often databases will be accompanied by a peer-reviewed paper that describes the data sources and how they have been manipulated. Databases with poor metadata and no such statements about their strengths and weaknesses have lower credibility.

Assessments are themselves prolific sources of data. Ensure that the assessment, in leaving behind legacy data, passes the tests just described. All the information used should be copied onto a secure medium after the assessment and placed in an archive that will outlive the project. Putting data into platform and software-independent formats is helpful where it may only be retrieved decades from now.

Make sure the metadata are adequate and, where possible, are embedded in the data files themselves, as self-describing column headings and units (see the simple guide to preparing datasets for sharing and archive by Cook et al. 2001).

Spatial data

Maps and remotely sensed images are valuable sources of information and potentially powerful communication tools. There are well-developed standards for documenting the origin and processing of such data, which should be adhered to both in the report and (in greater detail) in any electronically archived datasets (e.g., Moellering 2005).

There are trade-offs to be made with remotely sensed data between scale, resolution, content detail, frequency of acquisition, and cost. Scale is the extent of the coverage, and resolution is the smallest object that can be discerned, usually described by the pixel size. In general, large scales are associated with lower (coarser) resolution but also with higher frequency of acquisition and lower cost. The scale, resolution, and frequency need to be appropriate for the pattern expected in the image, which ultimately relates to the purpose of the study and the scale of the processes underlying the generation and use of ecosystem services. Higher spatial resolution and more spectral channels do not necessarily equate to "better" data. Remember that the cost of using remotely sensed information is not just the cost of acquiring the data but also the cost of analyzing it.

The usefulness of the information depends on both the content it represents and the level of processing it has undergone. Is it just one band, or is it multiple spectral bands? Are the bands in a part of the wavelength that contains the information being looked for? How well have the images have been corrected for geographical and atmospheric distortions? Uninterpreted images have limited use in assessments except as backdrops to maps or to show certain very obvious features such as large-scale land transformation. The most useful products are not the raw images themselves but products derived from them that relate directly to ecosystem services supply and use. For example, the São Paulo sub-global assessment used a Landsat 5 thermal infra red image (band 6) to illustrate the temperature-ameliorating service that green belts provide in urban heat islands. Land cover maps, tree cover maps, productivity maps, digital elevation models, and fine-resolution maps of the distribution of people are all derived from remote sensing. Such products are usually accompanied by a publication that describes how they were derived and validated and what the reliability of the product is.

Models

The distinction between "models" and "observations" is not nearly as sharp as many people think. Almost all measurements have some underlying model, and modern information sources such as satellite images have very elaborate models as part of the processing chain. Models are very useful in assessment because they allow gaps in space and time to be filled in a consistent way and they permit extrapolation, within reasonable limits. The citation for models must include the version used and should include a citation of the data used as input to the model. Many models are not truly in the public domain, and the same rule applies as for "gray literature"—a copy of the code needs to be placed in a public archive if it is to be used in an assessment.

Models that have passed peer review may be used in an assessment to conduct new analyses that do not need peer review (other than that provided by the assessment itself). This is the one form of "new research" that often is part of an assessment. For example, the Intergovernmental Panel on Climate Change commissions dedicated runs of many climate models for its assessments.

Just as with databases, it is important to assess and communicate the level of confidence in the models. This can be done by checking the quality of their documentation, by comparing them to other models or independent observations, or by any of the standard techniques of model validation. Models that do not permit any form of uncertainty analysis (i.e., where no confidence limit can be given) should be treated with caution.

Experiential, traditional, and indigenous knowledge

"Indigenous technical knowledge" and "traditional knowledge" are two of the terms used for information that may be well established and highly relevant to the assessment but that is not typically considered "scientific information." Assessments are encouraged to use such information, provided the rules of traceability and testability are satisfied. In addition, ethical rules relating to acknowledgement and use of intellectual property must be observed, and cultural sensitivities regarding privileged information must be respected. But if information from any source is so secret that it cannot be shared, it should not be offered to an assessment process nor used by such a process. The issues and approaches relating to use of traditional knowledge and similar sources are dealt with in more detail in Chapter 1.

Internet sources

Increasingly information is sourced from the Internet. Be aware that such information may not have passed a peer-review test and can be transient (i.e., untraceable in the future). It is also important not to assume that if information is not revealed by an Internet search, it must not exist. This discriminates against pre-Internet knowledge and knowledge in parts of the world with poor connectivity.

These caveats do not apply equally to all Web-accessible sources. There is no difference in credibility between an electronically published peer-reviewed journal and a traditionally published one, and the same applies to electronic databases that are supported by metadata and adequate descriptions of their origin. There is some evidence that open-source knowledge communities (such as Wikipedia) are of similar reliability as encyclopedia articles sourced from commissioned experts. Citizen–science-derived, Internet-based information systems (for instance, of bird observations) have proved very useful, especially if they have a built-in mechanism for quality checks.

4.2.3 Fitting biodiversity into an ecosystem service framework

The ecosystem services approach is based on a "utilitarian" concept of nature. In other words, nature is valuable because of its usefulness to humans. This is a defensible argument from the perspective that the concept of "value" is, as far as is known, an entirely human one, and it can only be conceived of from a human perspective.

Even humans' aesthetic, spiritual, cultural, or ethical appreciation of nature is, in this sense, "utilitarian." However, almost all researchers working in this field would concede that if humans were not present, nature would nevertheless have a "value." This is its "intrinsic value," in the strictest sense. (Some people use the term *intrinsic value* for the hard-to-monetize, nonconsumptive aesthetic, cultural, ethical, and spiritual values.) So an approach like that in the Millennium Ecosystem Assessment (MA) does not dismiss the existence of intrinsic value; it simply says that it is (by definition) unable to quantify intrinsic value and therefore cannot assess it. In this sense, utilitarian value is complementary to nonutilitarian value; not a replacement for it.

Biodiversity is in general not an ecosystem service itself but a necessary condition for ecosystem services to be delivered. Yet there are some important exceptions where the "diversity" part of biodiversity is itself the service—for instance, in some types of nature-based tourism and in the prospecting for new pharmaceuticals, genetic resources, and nature-inspired products. But in most cases of natural resource use, it is the "bio" part of biodiversity that is the service—as in the thousands of situations where a single particular species provides a food, medicine, or useful product. Here the product can be valued, but that does not itself constitute a value for biodiversity. Another species could provide the same or similar product, and often biodiversity is eliminated by the monocultural propagation of the preferred species. When biodiversity is one or two steps removed from the service—for instance, in most regulating and supporting services—attempts to value biodiversity directly are pointless and amount to double accounting. The economic value of biodiversity is embedded in the value of the services. This view (see Figure 4.1) represents an elaboration on the somewhat vague and simplistic representation of biodiversity (e.g., in the first versions of the MA conceptual framework, MA 2003) as a cloud of "life on earth" surrounding ecosystem services.

In general, the relationship between biodiversity and the delivery of ecosystem services is indirect. There is a large and growing body of evidence that the yield of many provisioning services is on average higher from systems with more biodiversity than from monocultures, although the maximum service yield from a monoculture may be equal to the maximum yield from a mixed system. The relationship between ecosystem service quantity and biodiversity approaches a maximum at a fairly low level of biodiversity (in the order of 10 species rather than hundreds). The relationship is clearer when biodiversity is expressed in terms of the number of functional types rather than the number of species.

A key value of biodiversity may be in reducing the variability of ecosystem services, (equivalently, reducing the uncertainty or risk), especially in the face of disturbances or changes in the environment. In other words, the value of biodiversity is expressed less via the average yield of the service (μ in figure 4.1) than via its higher moments, such as the variance (σ) or spatial diversity (γ).

Biodiversity can be measured and expressed in any number of ways (see a comprehensive treatment by Magurran 2004). Some of the key approaches are listed in Table 4.2.

4.2.4 Dealing with multiple scales

Assessments are typically carried out at particular scales: local, national, regional, or global, to name a few common ones. But they often rely on information collected at

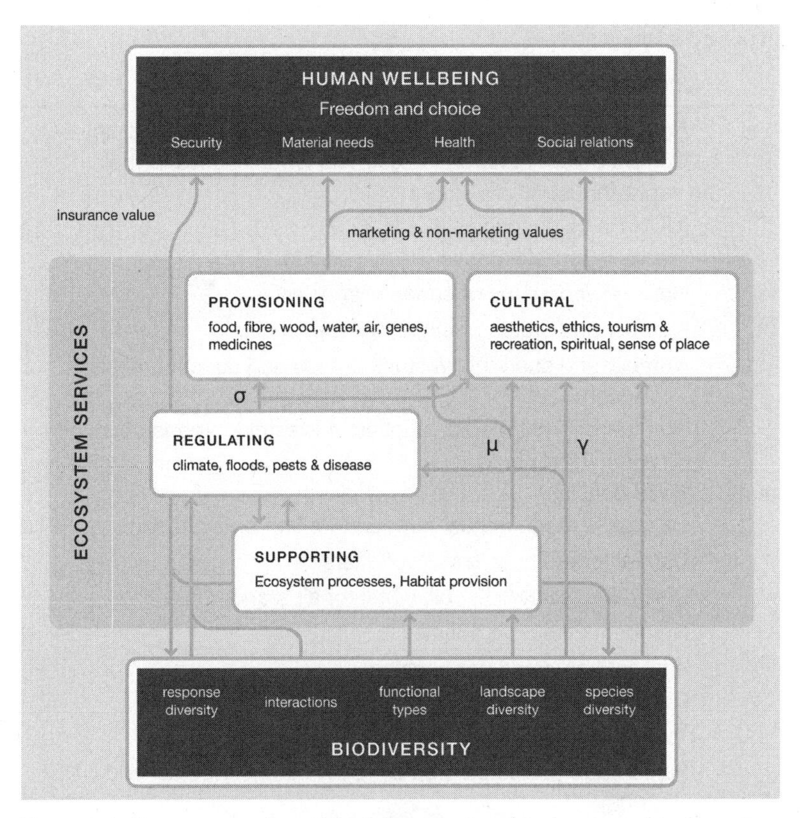

Figure 4.1. A conceptual model of the relationship between biodiversity and human well-being, via ecosystem services. The symbol μ indicates that the influence operates mostly through mean amount of the service that is generated, and that matters most in this influence, whereas σ indicates that the influence on temporal variability is the key factor, and γ indicates that the influence is mostly on spatial heterogeneity.
Source: Unpublished work by R. J. Scholes, C. Perrings and A. Kinzig.

greater or smaller scales, or they must interface with assessments or decision-making processes at somewhat different scales, either larger or smaller. The characteristic scale at which ecosystem services, their drivers, or the decision-making processes that influence them operate is seldom exactly the scale of the assessment. Since ecosystem services are place and time specific and do not necessarily aggregate upward in a simple additive way or disaggregate downward by simple proportionality, care must be taken in relation to scale questions in both time and space. It is important to work at an appropriate scale for both ecological and human processes, right from the start. (For a theoretical review of scale-related issues pertinent to ecosystem assessments, see MA 2003, chapter 7.)

As noted earlier, it is important to not confuse scale (the maximum extent or, in the time dimension, duration) with resolution (the smallest event detectable at that scale, sometimes called the "grain"). Usually the problems are caused by resolution mismatches rather than scale itself. Even fine-scale processes can be adequately analyzed at large scales, if the resolution is appropriate and there is enough computing power.

Table 4.2. Selection of biodiversity indicators

Aspect	Indicators
Ecosystem-level	Area of ecosystem types (e.g., forest, agriculture, built-up)
	Current extent of natural vegetation types relative to their pre-industrial era extent
Species-level	IUCN Red List species (IUCN 2002)
	Trends in representative or key species
	Number and status of endemic species
	Number and status of migratory species
Gene-level	Number and share of livestock breeds and agricultural plant varieties
	Number of endangered varieties of livestock breeds and agricultural crops
Pressures and threats	Road density
	Change in mean annual temperature and precipitation
	Damming and canalization of rivers
	Pollutants exceeding soil, water, or air standards
	Number of invasive species
Use	Amount harvested per species
	Carbon stored in forests
	Total revenue from ecotourism
Response	Percentage protected area by IUCN category (IUCN and UNEP 2003)
	Number of threatened and invasive species with a management plan
	Conservation policy capacity, in number of people
	Number of local site support groups and volunteer monitors
	Number of physical and chemical standards
Composite indicators	Hotspots (high endemism with high human impact) (Myers et al. 2000)
	Living Planet Index (Loh 2002)
	Natural Capital Index (Ten Brink 2000)
	Biodiversity Intactness Index (Scholes and Biggs 2005)
	Human footprint (Sanderson et al. 2002)
	Total Pressure Index (UNEP 2002)

Source: Based on Biggs et al. 2007, CBD 2003.

Provisioning services can generally be aggregated to higher scales if they are expressed in consistent and absolute units. Conservation of mass means that the supply of a mass or volume-based quantity of a service at a larger scale is the sum of the services at subsidiary scales, provided there is an unbiased sampling scheme. But in general even provisioning services expressed at a coarse resolution cannot be disaggregated into finer resolved form without using some form of probabilistic model. If the underlying drivers of the spatial or temporal variation in the service are known, and there is a representation of the fine-resolution variability (in time or space) of the driver, the driver field can be used as a covariate to predict a likely distribution. This process is used, for instance, to derive highly disaggregated population data from national census numbers, using information such as the distribution of night-time lights or roads as predictor variables.

The major problems arise for indicators where these "simple" rules do not apply. For instance, many biodiversity measures are scale dependent. An example is species richness: there is a strongly nonlinear relation between the number of species and the area under consideration. If the scale relationship is known, scaling laws can be applied. For example, in the species richness case, the scaling relationship is called the species-area curve, and it usually follows an exponentially saturating form with ecosystem-specific coefficients. When nonlinear scaling relationships are involved, the basic rule is that averaging (which is a form of aggregation) must take place at the end of the calculation rather than the beginning. In other words, the result, not the drivers, get aggregated.

A related problem is preserving the detail in findings when determining averages. Information often needs to be averaged in order to provide data summaries that are comprehensible. But the real world is invariably patchy, and there is a risk of homogenizing this spatial, temporal, and social variability in the average. It is possible to conclude, for instance, that there is no poverty at all because the average income is above some specified level, when in fact a large proportion of the population is very poor, but that is balanced by a small number of very rich individuals. One solution to this loss of information is to present the distribution of the values as well as the mean. This can take several forms, here listed in increasing levels of complexity: give the range, the standard deviation, higher order moments such as skewness and kurtosis, or the quantiles of the distribution. A graphic (for instance, a box-and-whisker plot) often captures this information more clearly than a table with lots of numbers. Spatial patchiness can be represented by color-shaded maps, while temporal variability can be shown by time-course graphics.

4.3 How to gather and assess information on ecosystem services

---

Section's take-home messages

- It is essential to be clear about whether an ecological stock or an ecosystem service flow is being measured. In general, stocks are expressed in units of quantity, while flows are expressed as quantities per unit time. Usually, ecosystem services are flows, both on the supply side and the demand side.
- Ecosystem integrity can be expressed through a combination of various indicators: the extent of the ecosystem relative to some reference state, the degree of fragmentation of remaining patches, the change in community composition of the ecosystem, and the capacity of the ecosystem to deliver a given service or basket of services per unit area.
- The "ecological footprint" indicator converts demand for a range of ecosystem services into the equivalent land or sea areas needed to supply that service and then adds them up. It can be calculated for an individual, an enterprise, or a nation.

This section describes the techniques used to collect and evaluate information about ecosystem services and the metrics used to express them. Table 4.3 provides a list of

Table 4.3. Some indicators and possible proxies for the main ecosystem services assessed in MA-type assessments

Class	Ecosystem service	Potential Indicator	Possible proxies
Provisioning	Food crops	Yield of crop product	Area planted to crop
	Livestock production	Offtake of animals or their products	Turnover or gross profit in meat, dairy and hide sectors
	Livestock as assets, draught animals, or cultural icons	Livestock biomass or metabolic equivalent mass (e.g., tropical livestock unit)	Livestock numbers by species
	Wild-harvested food (fisheries, hunting)	Offtake of given species Stocks of species	Turnover or gross profit of fisheries or hunting sector
	Energy crops, including fuelwood and charcoal	Yield (MJ) of given primary or secondary energy product	% of biofuels in energy mix
	Fiber (cotton, hemp, wool, silk, paper pulp, etc.)	Yield of given product (tons)	Turnover or gross profit of textile sector or paper-making industries
	Wood as timber	Harvest of products, usually as m^3, but also in local units such as board-feet or number of poles	Turnover or gross profit of forestry sector
	Fresh water	m^3 of fit-to-use water (for large flows, km^3 are used)	Per capita water use Price of water Cost of water purification Depth to groundwater
	Medicines	Harvest of known medicinal species (tons, or number of organisms)	Number of people using natural medicines
Regulating	CO_2 sequestration	Net CO_2 flux out of atmosphere	Change in C stock
	N,P, and S removal	Denitrification, P fixation, S precipitation	Downstream NO_3, PO_4 and SO_4
	Detoxification of waste	Difference in concentration of toxin in input and output stream	Illnesses attributable to toxins, incidence of fish kills
	Flood attenuation	Height and duration of flood peak	Losses of life and property due to flooding
	Coastal protection	Attenuation of coastal flooding, erosion and damage to infrastructure or resources	Km of coast with intact vegetation Cost of coastal damage
	Pest, pathogen and weed control	Intensity, duration, and extent of outbreaks of undesirable species	Expenditure on biocides Area occupied by alien species Number of alien species

Table 4.3. continued

Class	Ecosystem service	Potential Indicator	Possible proxies
Cultural	Recreation & amenity	Recreational opportunities provided	Tourism sector turnover or gross profit, number of visitors
	Aesthetic	Area of landscape in attractive condition	Visitor opinion polls Visits to beauty spots
	Spiritual and cultural	Presence of sites, landscapes, or species of spiritual or cultural significance	Number or area of important sites, protection status
	Scientific and educational	Presence or area of sites or species of scientific or educational value	Number of school visits Number of papers published
Supporting	Energy capture	Net Primary productivity	ΣNDVI, ΣFAPAR
	Nutrient cycling	N mineralization P mineralization Cation availability	Cover by N-fixing plants % mycorrhizae CEC in the profile, % base saturation
	Pollination	% of flowers pollinated within a species	Populations of pollinating species
	Habitat	Area of suitable habitat for a given species	Vegetation type area Fragmentation indices

ecosystem services and some of the indicators that have been found to be appropriate for quantifying them. The list is indicative rather than exhaustive; it covers the main, widely reported services. There are many more potential ecosystem services—sub-global assessments, for example, may reveal new ones—and each has many possible indicators and proxies.

4.3.1 Stocks and flows

It is essential to be clear about whether an ecological stock or an ecosystem service flow is being measured. For example, the amount of tree biomass in a forest is a stock; it supports several potential ecosystem service flows, such as the annual harvest of wood or an annual uptake of carbon dioxide. In general, stocks are expressed in units of quantity (e.g., metric tons, m^2, or ha), while flows are expressed as quantities per unit time (e.g., kg/year or m^3/second). Usually, ecosystem services are flows, both on the supply side and the demand side. Stocks and flows need to balance: if the demand exceeds the supply over a given period, the stock will be depleted by an equivalent amount. For a renewable resource, if the extraction rate is less than the natural replenishment rate, the stock will rise, usually to some maximum level.

It is usually necessary to express ecosystem services in both flow and underlying stock terms. The significance of a particular flow is hard to judge unless the size of the stock is known (and for renewable resources, the maximum flow that could be extracted from it without depleting the stock). Similarly, a stock by itself seldom says anything useful about the ecosystem service flows that are actually, or potentially could be, derived from it.

Both stocks and flows can be expressed in economic (monetary) terms or in physical quantity terms. The economic value of a flow is calculated as the quantity per unit time multiplied by an average price. The equivalent economic value of a stock that underlies a flow can be calculated from the present and assumed time course of future ecosystem service flows, using an appropriate discount rate (typically quite small, 1–3% per annum). This number is called the natural capital (Dasgupta and Maler 2001). Note that a stock or flow that is essential for survival and that cannot be substituted with another resource or with money would have an infinite value.

"Ecosystem services" are any benefit that people derive from nature, and they need not be restricted to living or renewable resources. Nonrenewable natural resources, such as ore bodies, fossil aquifers, and deposits of coal, oil, or gas, can also be regarded as natural capital stocks delivering a flow of services that end up supporting human well-being. For nonrenewable resources, the accounting must take into account their declining, nonself-replenishing nature—in other words, that the natural capital is depleted as the product is converted to financial or social capital. This must be factored into the net change in "inclusive wealth."

Not all ecosystem services are "consumed" when they are used. For instance, admiring a cultural landscape or a biodiversity icon does not necessarily make it unavailable to be admired by someone else. Even water is not destroyed when it is used: it is typically converted to another form (e.g., somewhat polluted) that may be unsuitable for immediate reuse for the same purpose but may be useful for another purpose.

The flows of provisioning services can often be directly measured, as a harvest yield over a period of time. Alternatively, they can sometimes be measured as a change in the stock over a given period. For instance, many ecosystems provide a climate regulating service by sequestering carbon from the atmosphere. It is possible to measure this flux of carbon dioxide (CO_2) from the atmosphere into the ecosystem directly, but the equipment needed is expensive and difficult to use. Over time, the net flux will show up as a change in the stock of carbon in the biomass, soil, sediment, or water body, and this is easier to measure.

4.3.2 Measures of ecosystem integrity

Ecosystem integrity, which is largely synonymous with ideas such as ecosystem health, quality, or intactness and is the converse of ideas such as degradation, can be expressed with combinations of the following indicators.

- The *extent of the ecosystem* relative to some reference state, such as its former extent at some time during the period of historical record, or the extent inferred from paleo-ecological data, or a "potential" distribution inferred from climate or substrate requirements. This is the simplest indicator of ecosystem loss, but it is a measure of stock rather than ecosystem service flow.
- The *degree of fragmentation* of the remaining patches, which can be expressed using a wide variety of metrics, including average patch size, the perimeter:area ratio, and the degree of connectivity between patches. These measures are particularly important with respect to the habitat-supporting service, since the number and type of species that can persist in a patch depends, in the long term, on the size of the patch and its exposure to disturbance along its perimeter.

- The *change in community composition* of the ecosystem, usually expressed relative to some reference state (such as the "natural state" inferred from some examples believed to be in this state). The difference metric must be able to cope with simultaneous change across many species. Some are quite simple (such as the Euclidean distance to the reference state in the "hypervolume" defined by the abundances of each species), while others use advanced statistical concepts, such as collapsing the many variables onto a few "principal components." Composition measures are very sensitive—perhaps oversensitive—to natural fluctuations and variations in species composition. They can be hard to interpret: is an observed change in composition a good or bad thing? An alternative approach is to reduce the community diversity to an index, such as the species richness or the Simpson's Diversity index, and then track changes in that. Note that the link between compositional diversity and ecosystem service delivery remains generally unclear.
- The *capacity of the ecosystem to deliver a given service*, or basket of services, per unit area. This is a sensitive and useful indicator and forms a robust and defensible basis for the definition of concepts such as degradation. The MA Desertification Synthesis defines degradation as a persistent reduction in the capacity of the land to deliver one or more ecosystem services. Service-based integrity indicators pick up the subtle changes in ecosystem functioning that are missed by simple loss-of-area measures—for instance, the high grading of valuable timber out of forests without clear-cutting or the consumption of all the bushmeat from under the canopy.

4.3.3 The balance of ecosystem service supply and use

Some intuitive ecosystem service indicators combine measures of ecosystem service demand (consumption) with estimates of the supply capacity. The simplest of these calculates the difference between supply and demand (see Box 4.3 for an example). This is a better approach than calculating a supply:demand ratio, which becomes undefined if either tends to zero.

A composite form of this kind of indicator is the ecological footprint. It converts demand for a range of ecosystem services into the equivalent land or sea areas needed to supply that service and then adds them up (Rees and Wackernagel 1994, see also http://www.footprintnetwork.org). The ecological footprint can be calculated for an individual, an enterprise, or a nation. A footprint is unsustainable when it is greater than the area of ecosystems that are equitably available to that entity. The ecological footprint is a very graphic and easily communicated way of expressing ecosystem service supply versus demand, but it has several problems. First, although all the components are expressed in the same units (hectares) and can therefore be mathematically added, it is not clear that they can be ecologically added, since often the same hectare can deliver several services simultaneously, or with only weak trade-offs between them. Second, the area needed to supply the service may not be based on the yield where the service is actually derived (which is almost impossible to trace or calculate) but on some "mean yield," which is easily calculated but may not be relevant.

Box 4.3. Calculation of the fuelwood deficit in Southern Africa

In the Southern Africa Millennium Ecosystem Assessment, the availability of fuelwood as an ecosystem service was assessed by calculating the difference between supply and demand. Fuelwood supply was calculated at a 5 × 5 km resolution based on climate and satellite-derived tree cover data. Fuelwood demand (b) was calculated by scaling average consumption in rural and urban areas by a function of ambient temperature and woodfuel availability. The results show clearly that, contrary to popular conception, woodfuel scarcity in southern Africa is confined to very specific parts of the region.

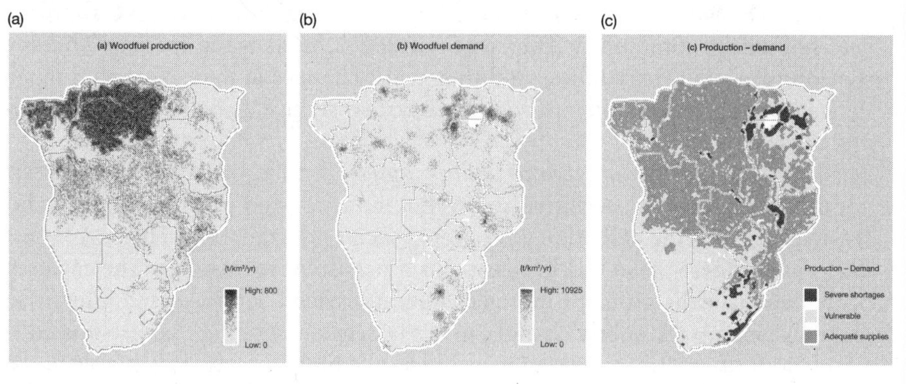

Box 4.3 Figure. Supply (a), demand (b) and shortfall or surplus (c) of fuelwood in southern Africa in 2000.

Source: Scholes and Biggs (2004)

4.4 How to gather and assess information on human well-being

Section's take-home messages

- In the MA, human well-being is considered to have many dimensions, including but not restricted to monetary income. In this view, *poverty* is defined as the absence of well-being.
- Values expressed in monetary units that are accrued over a multi-year period are corrected for the effects of inflation, and are often converted to an internationally comparable currency.

The MA takes a particular position on human well-being: that it is multidimensional and, in particular, includes considerations in addition to monetary income (see MA 2003, Chapter 3). In this view, *poverty* is defined as the absence of well-being rather than simply as not having a certain minimum amount of money. Some important axes of human well-being, and their indicators, are summarized in Table 4.4. (There are several comprehensive discussions of this topic; see Alkire 2002, Prescott-Allen 2001.)

Table 4.4. The constituents of human well-being recognized by the Millennium Assessment, and some potential indicators for them

HWB constituent	Subcategory	Potential Indicators
Basic material for a good life	Access to resources to sustain livelihood	Offtake of food, fiber, fuel, medicines, construction materials and freshwater to meet basic needs
		Security of resource or land tenure
	Income	Income (including from sale of above)
		Measures of income distribution
Health	Nutrition	Protein intake per day
		Digestible energy in food
		Deviation from target weight per height
	Disease	Expected longevity at birth
		Childhood mortality
		Disease-adjusted life years
	Exposure to toxins	Exceedance of guideline limits
		Prevalence of indicator health conditions
	Comfort	Ability to keep warm or cool
Security	Shelter	People in adequate housing
	Exposure to risk	Risk of death, injury, or property loss through natural hazards
Good social relations	Absence of conflict	Deaths, injuries, property, or infrastructure loss; number of displaced people due to armed conflict
	Sense of belonging	Happiness measures
Freedoms and choice	Participation in decision making	Level of education
		Gender bias in education
		Level of corruption
		Fairness of elections

The assessment team needs to make a number of decisions on technical issues. For measures that are expressed in monetary terms, income over a multiyear period is usually expressed as the equivalent at a given time, to allow for the effects of inflation. For instance, an assessment over the period 1991 to 2000 may choose to express its income in equivalent year 2000 terms. To do this, econometric time series of inflation indexes (for example, the consumer price index) are needed. Second, the income is often expressed not only in local currency terms but also in an internationally comparable currency, such as the U.S. dollar. To do this, a rate of exchange and a date must be specified. Often this rate of exchange is not based on the market rate but on the purchasing power parity (PPP) conversion rate, which is more reflective of the impact on human well-being.

Many indicators have the tendency to mask pockets of low well-being by averaging across the entire population. For instance, gross income indicators (such as the GDP/capita) by themselves are less informative than income plus some measure of its distribution in society: the quintiles of income, for example, or some composite measure of income equality, such as the Gini coefficient, or some measure such as the fraction of the population below a specified income limit. Similarly, health measures such as longevity or child mortality are more informative if they also give an indication of the distribution of the indicator in the population. They can sometimes be disaggregated spatially (since the primary data source is usually the local clinic). If possible, they should be expressed per vulnerable class (e.g., gender, ethnic group, income class, or age).

A number of composite indices of human well-being have been attempted, and some are routinely calculated and available at the national scale. The U.N. Development Programme (UNDP) annually reports the Human Development Index for every country. This combines three dimensions: health (measured by life expectancy at birth), education (measured by the adult literacy rate and the combined gross enrolment ratio across all educational levels), and income (measured by the logarithm of gross domestic product per capita, at purchasing power parity in dollars). UNDP's Human Poverty Index is a geometric average of three indicators: the probability of surviving to age 40, the fraction of adults who are literate, and the average of the fraction of the population that has access to safe water supplies and the fraction of children who are underweight for their age.

The *Wellbeing of Nations* prepared by Robert Prescott-Allen combines, for each country, 36 indicators of health, population, wealth, education, communication, freedom, peace, crime, and equity into a Human Wellbeing Index. It also combines 51 indicators of land health, protected areas, water quality, water supply, global atmosphere, air quality, species diversity, energy use, and resource pressures into an Ecosystem Wellbeing Index. The ratio of the two indexes is an index that measures how much human well-being each country obtains for the amount of stress it places on the environment (Prescott-Allen 2001).

The Millennium Development Goals adopted by the United Nations aim to raise the well-being of people all over the world. They are associated with specific targets, each of which has one or more indicators. Data on these indicators are often available at national resolution, and sometimes at sub-national resolution (see Table 4.5).

4.5 How to assess the link between ecosystem services and human well-being

---

Section's take-home messages

- The first step in establishing the link between an ecological resource or service and human benefit is to sketch out a causal pathway linking the service in question to the elements of human well-being it is thought to influence.
- Economic valuation of ecosystem services is a complex technical field that can be difficult for untrained and inexperienced people to master. Yet several relatively simple steps can lead in the direction of economic valuation, such as expressing the service flows and their underlying stocks in quantitative terms.
- It is important to calculate the economic value of ecosystem services—the all-inclusive value to society as a whole, over the entire life cycle, taking into account both taxes and subsidies and adding in the externalities—rather than just the financial value to the immediate beneficiary.
- A commonly used tool to assess trade-offs is the social preference function. Trade-offs occur when the extraction and use of one service has an impact—positive or negative—on the benefit that can be realized from another service.

Table 4.5. Indicators used to quantify progress toward the U.N. Millennium Development Goals

Goal and targets	Indicators
Eradicate extreme poverty and hunger. ***Target 1:*** Halve, between 1990 and 2015, the proportion of people whose income is less than one dollar a day. ***Target 2:*** Halve, between 1990 and 2015, the proportion of people who suffer from hunger.	**1.** Proportion of population below $1 (1993 PPP) per day (World Bank) **2.** Poverty gap ratio [incidence × depth of poverty] (World Bank) **3.** Share of poorest quintile in national consumption (World Bank) **4.** Prevalence of underweight children under five years of age (UNICEF–WHO) **5.** Proportion of population below minimum level of dietary energy consumption (FAO)
Achieve universal primary education. ***Target 3:*** Ensure that, by 2015, children everywhere, boys and girls alike, will be able to complete a full course of primary schooling.	**6.** Net enrollment ratio in primary education (UNESCO) **7.** Proportion of pupils starting grade 1 who reach grade 5 (UNESCO) **8.** Literacy rate of 15–24 year olds (UNESCO)
Promote gender equality and empower women. ***Target 4:*** Eliminate gender disparity in primary and secondary education, preferably by 2005, and in all levels of education no later than 2015.	**9.** Ratio of girls to boys in primary, secondary, and tertiary education (UNESCO) **10.** Ratio of literate women to men, 15–24 years old (UNESCO) **11.** Share of women in wage employment in the nonagricultural sector (ILO) **12.** Proportion of seats held by women in national parliament (IPU)
Reduce child mortality. ***Target 5:*** Reduce by two-thirds, between 1990 and 2015, the under-five mortality rate.	**13.** Under-five mortality rate (UNICEF–WHO) **14.** Infant mortality rate (UNICEF–WHO) **15.** Proportion of 1-year-old children immunized against measles (UNICEF–WHO)
Improve maternal health. ***Target 6:*** Reduce by three-quarters, between 1990 and 2015, the maternal mortality ratio.	**16.** Maternal mortality ratio (UNICEF–WHO) **17.** Proportion of births attended by skilled health personnel (UNICEF–WHO)
Combat HIV/AIDS, malaria, and other diseases. ***Target 7:*** Have halted by 2015 and begun to reverse the spread of HIV/AIDS. ***Target 8:*** Have halted by 2015 and begun to reverse the incidence of malaria and other major diseases.	**18.** HIV prevalence among pregnant women aged 15–24 years (UNAIDS–WHO–UNICEF) **19.** Condom use rate of the contraceptive prevalence rate (UN Population Division) **19a.** Condom use at last high-risk sex (UNICEF–WHO) **19b.** Percentage of population aged 15–24 years with comprehensive correct knowledge of HIV/AIDS (UNICEF–WHO) **19c.** Contraceptive prevalence rate (UN Population Division)

(Continued)

Table 4.5. continued

Goal and targets	Indicators
	20. Ratio of school attendance of orphans to school attendance of nonorphans aged 10–14 years (UNICEF–UNAIDS–WHO) **21.** Prevalence and death rates associated with malaria (WHO) **22.** Proportion of population in malaria-risk areas using effective malaria prevention and treatment measures (UNICEF–WHO) **23.** Prevalence and death rates associated with tuberculosis (WHO) **24.** Proportion of tuberculosis cases detected and cured under DOTS (internationally recommended TB control strategy) (WHO)
Ensure environmental sustainability. ***Target 9:*** Integrate the principles of sustainable development into country policies and programs and reverse the loss of environmental resources. ***Target 10:*** Halve, by 2015, the proportion of people without sustainable access to safe drinking water and sanitation. ***Target 11:*** By 2020, achieve a significant improvement in the lives of at least 100 million slum dwellers.	**25.** Proportion of land area covered by forest (FAO) **26.** Ratio of area protected to maintain biological diversity to surface area (UNEP–WCMC) **27.** Energy use (kg oil equivalent) per $1 GDP (PPP) (IEA, World Bank) **28.** Carbon dioxide emissions per capita (UNFCCC, UNSD) and consumption of ozone-depleting CFCs (ODP tons) (UNEP–Ozone Secretariat) **29.** Proportion of population using solid fuels (WHO) **30.** Proportion of population with sustainable access to an improved water source, urban and rural (UNICEF–WHO) **31.** Proportion of population with access to improved sanitation, urban and rural (UNICEF–WHO) **32.** Proportion of households with access to secure tenure (UN–HABITAT)
Develop a global partnership for development. ***Target 12:*** Develop further an open, rule-based, predictable, nondiscriminatory trading and financial system. ***Target 13:*** Address the special needs of the least developed countries (includes tariff and quota-free	**33.** Net ODA, total and to LDCs, as percentage of OECD/Development Assistance Committee (DAC) donors' gross national income (GNI) (OECD) **34.** Proportion of total bilateral, sector-allocable ODA of OECD/DAC donors to basic social services (basic education, primary health care, nutrition, safe water, and sanitation) (OECD)

Table 4.5. continued

Goal and targets	Indicators
access for least developed countries' exports, enhanced program of debt relief for heavily indebted poor countries (HIPC) and cancellation of official bilateral debt, and more generous ODA for countries committed to poverty reduction). ***Target 14:*** Address the special needs of landlocked developing countries and small island developing states (through the Programme of Action for the Sustainable Development of Small Island Developing States and the outcome of the twenty-second special session of the General Assembly). ***Target 15:*** Deal comprehensively with the debt problems of developing countries through national and international measures in order to make debt sustainable in the long term. ***Target 16:*** In cooperation with developing countries, develop and implement strategies for decent and productive work for youth. ***Target 17:*** In cooperation with pharmaceutical companies, provide access to affordable essential drugs in developing countries. ***Target 18:*** In cooperation with the private sector, make available the benefits of new technologies, especially information and communications.	**35.** Proportion of bilateral ODA of OECD/DAC donors that is united (OECD) **36.** ODA received in landlocked developing countries as a proportion of their GNIs (OECD) **37.** ODA received in small island developing states as proportion of their GNIs (OECD) **38.** Proportion of total developed-country imports (by value and excluding arms) from developing countries and from LDCs, admitted free of duty (UNCTAD, WTO, World Bank) **39.** Average tariffs imposed by developed countries on agricultural products and textiles and clothing from developing countries (UNCTAD, WTO, World Bank) **40.** Agricultural support estimate for OECD countries as percentage of their GDP (OECD) **41.** Proportion of ODA provided to help build trade capacity (OECD, WTO) **42.** Total number of countries that have reached their HIPC decision points and number that have reached their HIPC completion points (cumulative) (IMF, World Bank) **43.** Debt relief committed under HIPC initiative (IMF, World Bank) **44.** Debt service as a percentage of exports of goods and services (IMF, World Bank) **45.** Unemployment rate of young people aged 15–24 years, each sex and total (ILO) **46.** Proportion of population with access to affordable essential drugs on a sustainable basis (WHO) **47.** Telephone lines and cellular subscribers per 100 population (ITU) **48.** Personal computers in use per 100 population and Internet users per 100 population (ITU)

Source: based on UNGA 2001.

Ecosystem services support human well-being via many (but not infinitely many) paths, as indicated earlier in Figure 4.1. The crucial first step in establishing the link between an ecological resource or service and human benefit is to sketch out a causal pathway, which can be thought of as a hypothesis, linking the service in question to the elements of human well-being it is thought to influence. This pathway may have several steps, especially if the service in question is a supporting or regulating service. Quantifying the key elements in this pathway is the next step, and this will rely on appropriate indicators at each step of the pathway. These indicators may be derived from observations, from models, or (most likely) from some combination of both.

It is important to bear in mind that human well-being is multidimensional and is influenced by factors that may be unrelated to ecosystem services. For instance, growth of the manufacturing economy or the availability of foreign aid may increase well-being, but they are very indirectly and distantly related to ecosystem services. Furthermore, impacts on human well-being may be slow to materialize. Therefore it will often be hard to unequivocally demonstrate that a particular service has been solely responsible for an observed change in human well-being. This is not a problem unique to ecosystem services. It is an intrinsic feature of impact analysis in highly connected environments.

4.5.1 Ecosystem service valuation

The economic valuation of ecosystem services is not essential to an ecosystem service assessment, but when it is possible to do one in a reasonably rigorous way, it is very useful for three reasons:

- The process imposes a high level of rigor on the assessment.
- Being able to express different services in a common denominator (economic value) allows trade-offs to be explicitly evaluated.
- Communication of the importance of ecosystem services within a policy environment is greatly facilitated if it is accompanied by credible economic values.

A word of caution is needed. Economic valuation of ecosystem services is a complex technical field that can be difficult for untrained and inexperienced people to master. Often, such evaluations do not already exist when an assessment is launched, so the assessment would need to undertake the studies itself: a violation of the guideline that assessments should not engage in new research. Furthermore, such studies may be costly and time consuming. Only venture into economic valuation as far as your team's skills, funds, and time allow.

On a more positive note, several relatively simple steps can lead in the direction of economic valuation and are very useful. For instance, expressing the service flows and their underlying stocks in quantitative terms is an important step. Describing the way in which they lead to human benefits is another key step, even if it cannot be quantified. Often it is relatively straightforward to calculate some of the components of total economic value, even if they cannot all be estimated—and this may be enough to make the point that ecosystem services are valuable. For instance, use values are often easier to estimate than nonuse values, and within use values, actual direct use values are easier to estimate than indirect use values or option values.

Economists are quick to point out that economic value is not synonymous with monetary value. Money is simply a convenient way of expressing the value that society applies to things. Some values can be denominated in other terms, which may be more appropriate than money in certain cases—the number of lives lost, for instance. For the purposes of ecosystem services assessment, it is the economic value that matters (i.e., the all-inclusive value to society as a whole, over the entire life cycle, taking into account both taxes and subsidies and adding in the externalities) rather than the financial value (which is calculated at the level of the entity being accounted only, excludes costs and benefits relating to the broader context, and is expressed only in monetary terms). Many things that are financially profitable are economically loss making, and vice versa.

Some ecosystem service assessments calculate the total value of the service (e.g., Costanza et al. 1997). Note that "total value" can have several meanings. In contrast to marginal value, it means the accumulated value to all beneficiaries of the total consumption of a single service, over the entire area and duration of the assessment. Total economic value can also mean the value accrued by all the mechanisms illustrated in Figure 4.2. Total value could also mean the value accumulated across all the different ecosystem services that exist in an area or just the subset of services for which value can be calculated. The phrase "global value" is also sometimes used as a synonym to total value, which gets confusing in the context of global versus sub-global ecosystem assessments, so we do not recommend using it.

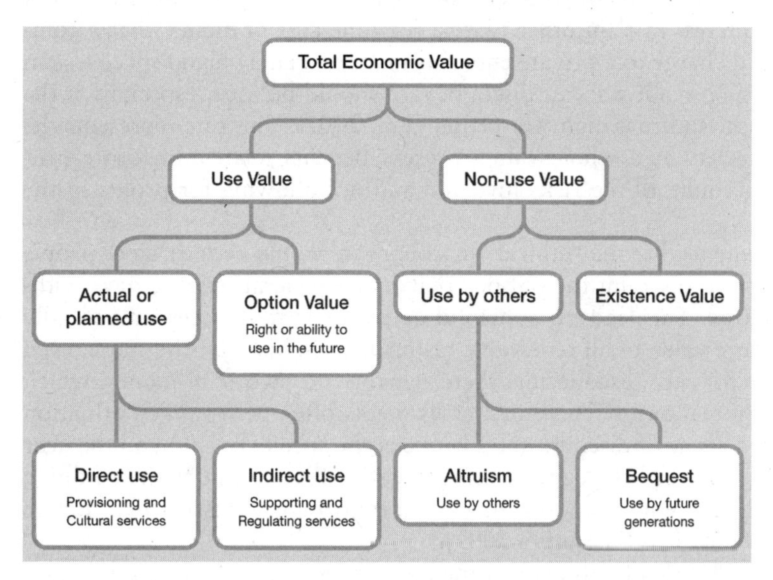

Figure 4.2. The Total Economic Value classification of the components of value of ecosystem services Note that "non-use value" should not be confused with "non-consumptive use"—i.e., use that does not reduce the stock. It means that the person who is assigning value is not the person who uses the service. Further, "existence value" should not be confused with "intrinsic value" (the ethically based value of other species to exist, independent of the benefits they may yield to humans). If humans can value it, however esoterically, it is by definition an extrinsic value. Note that in the indirect use values should be completely reflected in the direct values and should not be counted twice.
Source: Adapted from DEFRA 2007

Usually the more useful value to calculate is the marginal value—in other words, the extra value that is added by one more unit of the service over and above the current supply. This is more applicable for policy purposes where small adjustments in the balance between services are envisioned rather than the total replacement of a service. The marginal value is the relevant information for calculating the trade-off with other options within a small range on either side of the current state. Marginal values are calculated by well-established methods that typically depend on solving economic equilibrium models.

Often the value of an ecosystem service is equated to the financial turnover of that commodity—in other words, the volume multiplied by the price. This is sometimes all that is possible to calculate, and it is permissible if it is used in a comparative sense, in relation to values for other services or economic activities that are calculated in the same way, as a sort of relative index of importance. Strictly speaking, it is the value addition of a service that should be calculated—that is, the realized income from that service less the cost of production. For instance, the value of the provisioning service provided by a particular crop is not the market price multiplied by the yield; it is this price less the cost of the inputs needed to achieve the yield and the cost of externalities (the costs borne by society at large, such as pollution of a river). It should also be corrected for subsidies.

A technical question that always arises is the appropriate discount rate to use for bringing future costs and benefit flows to present terms (see European Communities 2008, Chapter 3). There is no consensus on this issue, but general agreement that a fully commercial interest rate (in other words, the time cost of money that a commercial bank would charge to a private entity) is inappropriately high for ecosystem service assessments. Instead, a social discount rate should be used, especially if the uncertainty about the future is high (Carpenter et al. 2007). This rate represents the degree to which society as a whole values present benefits relative to future benefits, taking into account all the risks involved and not allowing for profits in the transaction.

There are arguments that the "ethical" discount rate would be zero, since people today have no right to discount the options that future generations can have without their participation. This leads to unhelpful outcomes in most cases—it basically means that the future value of all renewable resources is infinite. Furthermore, even if the ethics argues for zero discounting, there remains a time cost of money, which may be quite low in real terms. Therefore, assessments often perform key valuation calculations using a range of discount rates between the social (1–3 %) and the near market (10–15%). Some work suggests using negative discount rates (i.e., valuing the future above the present) in addressing crucial environmental issues like climate change and biodiversity loss (Dasgupta 2007).

There are many detailed treatments of the techniques of ecosystem service valuation. (A useful guide in the context of ecosystem services is CBD 2007; in the same Technical Series, also see TS 3 on wetlands, TS 4 on forests, and other relevant titles. See also DEFRA 2007.) The techniques rely either on "revealed preference" (actual choices made in real markets, usually only applicable to use values) or on "stated preferences," based on hypothetical choices elicited through questionnaires. Revealed preference methods are usually based on "willingness to pay," whereas stated preferences may be based either on willingness to pay for a service or willingness to accept payment for a service. Table 4.6 gives an overview of the techniques that may

Table 4.6. Methods of ecosystem service valuation

Technique	Brief explanation	Advantages	Problems
Production function	What is the impact on the production of a marketed good caused by a change in the supply of an ecosystem service	Linked to actual markets	Needs a clear and quantified understanding of the causal relationship between the service and the product
Market value	Price times volume gives the value	Data readily available and reliable	Applicable to directly marketed ecosystem services only (i.e., provisioning services and some cultural services)
Cost-based (also called shadow pricing, substitution pricing)	Usually calculated as what someone would have to pay to get the service in another way or to restore a service if it were lost or damaged		Can overestimate the actual value
Cost of illness or morbidity (also called human capital approach)	Calculates the equivalent value of the loss of earnings or life that would result if the service were not effective	Applicable to regulating services such as the absorption of toxins and pollutants or the control of diseases	Controversial when applied to mortality using inequitable "value of life"
Hedonic pricing	Value that shows up in the prices of other goods, for instance the increase in property value where the supply of a service is high	Linked to markets, so data are available and reliable	Data intensive, and mostly related to property prices
Contingent valuation and choice modeling	Based on questions such as "what would you be willing to pay for this service?"	Applicable to all ecosystem services, and to use and nonuse values	Subject to bias and high uncertainty
Travel cost	What people are shown to be willing to pay to travel to destinations where cultural services (beauty, recreation) are on offer	Based on observed behavior	Limited to some cultural services; can overestimate if a trip is made for many purposes
Random utility	Extension of travel cost method; examines the impact of changing quality or quantity of an ecosystem service at a given site		Limited to use values, mostly for recreational services

be useful in particular situations. (For more details, references, and examples, see CBD 2007 and Defra 2007.)

In some circumstances nonmonetary valuation is more appropriate than monetary valuation—for instance, in barter economies. It may be ethically inappropriate to place a value on human life because such values often implicitly devalue the lives of the poor relative to the rich. In these cases a variety of techniques can be used—for instance, preference ranking for certain services, undertaken using participatory methods (Howarth and Wilsdon 2006), or measurement in terms of other human well-being indicators, such as life span, disease-adjusted life years, infant mortality, or weight-for-age curves.

4.5.2 Assessing trade-offs

A trade-off occurs when the extraction and use of one service has an impact on the benefit that can be realized from another service (Scholes 2009). This impact can be positive or negative. A positive interaction is called a synergy or, in popular terms, a win-win situation. For instance, many of the supporting, regulating, and cultural ecosystem services are synergistic—actions taken to strengthen one also strengthen others.

Unfortunately, the more common and worrying case is a negative trade-off (a win-lose or lose-lose situation). For instance, increasing the production of one provisioning service usually reduces the availability of other provisioning services and of regulating, supporting, and cultural services. The issue of assessing trade-offs is discussed further in Chapter 6. Some of the methods that can be used are very sophisticated. If these are beyond the assessment's capacity, at least show which services are likely to be involved in trade-offs, and try to give an indication of the nature of the trade-off (positive or negative) and its approximate magnitude, in a relatively simple matrix form (see, for example, Table 4.7).

A commonly used tool to assess trade-offs is the social preference function. For individuals, this reflects the preference ordering of a range of constituents of well-being that the individual has reason to value. The list of constituents as well as their weights will differ among individuals. The preference ordering provides information on the perceived relative importance of various ecosystem services as determinants of the various constituents of human well-being. A commonly used approach to solicit the social preference ordering is through deliberative participatory techniques, whereby groups of individuals are asked to rank their preferences over a range of constituents of well-being with and without the knowledge of the links between the various constituents of well-being and ecosystem services.

4.6 How to communicate state and trends

Section's take-home messages

- Assessments are a communication tool between researchers and decision makers. If they are technically proficient but fail to communicate, they fail overall.
- For many issues there are vast amounts of data and information available. Having specific questions to be answered will guide selection of the appropriate data and the techniques for collating and synthesizing the data.

Table 4.7. An example of a trade-off matrix. This one evaluates divergent potential uses of a tropical rainforest ecosystem.

	Global environmental concerns		Agronomic sustainability (0 = no issue, 1 = major issue)			National policy makers' concerns		Smallholders' concerns/Adoptability by smallholders	
Land-use systems	Carbon storage	Biodiversity	Plot-level production sustainability			Potential profitability[b]	Labor requirements	Returns to labor[‡]	Household food security[c]
	Above-ground t C ha^{-1} (time averaged)[a]	Species per plot	Soil structure	Nutrient export	Crop protection	Returns to land (private prices) R$ ha^{-1}	Labor person-d $ha^{-1}yr^{-1}$	$/person-d (private prices)	Entitlement path (operational phase)
Forest	148	80	0	0	0	–2	1	1	NA
Managed forestry	~148	NM	0	0	0	416	1.22	20	$
Coffee/*Bandarra*	56	27	0.5	0.5	0.5	1955	27	13	$
Coffee/Rubber	56	16	0.5	0.5	0.5	872	59	9	$
Traditional pasture	3	10	0 to 1	0.5	0.5 to 1	2	11	7	$ + consumption
Improved pasture	3	NM	0 to 1	0.5	0.5 to 1	710	13	22	$ + consumption
Annual/Fallow	7	34	0 to 0.5	0 to 0.5	0.5 to 1	17	23	6	$ + consumption
Improved fallow	~3–6	26	0 to 0.5	0 to 0.5	0.5 to 1	2056	21	17	$ + consumption

NM = not measured; NA = not applicable.

[a] Indicates time averaged above-ground carbon.

[b] Prices are based on 1996 averages, and expressed in December, 1996 R$ (US$ = R$1.04), discounted at 9% per annum.

[c] *Food Security*, *consumption*, and $ reflect, respectively, whether the technology generates food for own consumption or income that can be used to buy food, or both.

Source: Tomich, T. P. et al. 2005, (Table 1).

- Assessments should report variation in ecosystem services over space and time. It is often the variability in the production of the service that most affects human well-being rather than the mean availability.
- It is fundamental to good scientific practice, and especially important in the context of assessments, to accompany key assertions with some measure of the confidence in those findings. For quantitative data, a statistical approach is possible; for more qualitative findings, a set of agreed phrases regarding the evidence and amount of agreement can be used.
- Care should be taken in presenting complex information in easily understood language and in different formats that are suitable for a variety of audiences.

Assessments are a communication tool between researchers and decision makers. If they are technically proficient but fail to communicate, they fail overall. Therefore a great deal of attention must be given to choosing the best ways of presenting the information to the intended audience and making the overall product readable, understandable, and unambiguous.

4.6.1 Collate, summarize, analyzes, and synthesize

If assessments do not collect new data, what value are they? They add value by bringing the data together, analyzing it, evaluating it, and showing connections and meaning that were not available in the raw data.

Any data collation or synthesis exercise should rest on careful consideration of what information is relevant and useful in terms of the particular issue being examined. Although collecting and analyzing data can be tedious, it should never be approached as a mindless task. Data should not be reported simply because they are available. Before collating and synthesizing the data, it is usually a good idea to brainstorm and test the questions the intended audience would like answered. For many issues there are vast amounts of data and information available. Having specific questions to be answered will guide selection of the appropriate data and the techniques for collating and synthesizing the data.

The methods used for summarizing data will depend on the information to be provided. Some of the most common types of information provided in assessments of ecosystem services are measures of "central tendency" (i.e., the mean, median, or mode, as appropriate) and measures of variability over space or time (i.e., the range and/or standard deviation). Analyzing how different variables relate to one another can be very informative, but it may require fairly sophisticated statistical techniques. For any information provided in an assessment, it is important to note the degree of confidence that can be associated with the information. This can take the form of formal statistics (e.g., a confidence interval) for quantitative data or language that conveys uncertainty for qualitative data (see next section).

Assessments often focus largely on reporting means, but they should probably pay as much or more attention to reporting variability in ecosystem services over space and time. It is often the variability in the production of the service that most affects human well-being rather than the mean availability. For instance, freshwater availability in the driest month of the year—not the average annual freshwater availability—is the critical constraint on agricultural production. The most common measures of variability or spread are the range (the difference between the largest and smallest value) and the standard deviation (a measure of the average difference between each individual data point and the mean of all data points). "Box and whisker plots" are a very nice graphic tool for summarizing the mean, standard deviation, and range of a particular dataset. Maps are powerful tools for displaying spatial variation in ecosystem services. When comparing variation among different variables, it is most appropriate to use the coefficient of variation, which measures the ratio of the standard deviation to the mean (usually as a percentage) and is therefore dimensionless.

Understanding relationships between different ecosystem services or between drivers of ecosystem change and ecosystem services can be very useful for assessing trade-offs between services or proposed management interventions. Relationships are often best conveyed graphically as biplots (i.e., the one value plotted against the other value, as points or as lines). The simplest statistical measure of a relationship between two or more variables is the correlation coefficient. But beware: a correlation between two variables does not provide proof of a causal relationship. Causal relationships, and predicting changes in one variable from a change in another, are typically examined using regression analysis, but regression does not "prove" a link either. It just creates a more plausible case. Regression analyses range from simple linear regression to highly sophisticated models with nonlinear relationships, many predictor variables, time delays, and causal hierarchies. Gelman and Hill (2007) provide good coverage of these techniques.

4.6.2 Conveying certainty and uncertainty

The purpose of assessments is not to fill knowledge gaps but to reveal them. It is important, therefore, not try to paper over the cracks but to be honest about what is not known and what is uncertain. Sometimes small missing pieces of information can be obtained during the assessment process, and small gaps can be estimated using extrapolation. But there will invariably still be information that is either completely missing or to some degree inadequate. An important function of assessments is to establish the research priorities for the future.

It is fundamental to good scientific practice, and especially important in the context of assessments, to accompany key assertions with some measure of the confidence in those findings. This obligation will not always be demanded by the audience, and it may even be resisted ("Can't these scientists make up their mind?"). Not every statement in the entire text needs to be accompanied by a qualification or an error range, but the key ones do. Be vigilant in eliminating overqualification of findings. For example "The following result may occur with a low probability..." is a redundancy.

Certainty (or uncertainty) can be presented several ways in assessments (Schneider et al. 1998, Moss and Schneider 2000, MA 2003:175–76). The method used depends on the type of information involved. For richly quantitative data, a statistical approach (in other words, the familiar confidence limits in tables, graph, or text) is possible and can be interpreted in text with special "reserved words" (see Box 4.4). For more qualitative findings, a set of agreed phrases can be used.

There are two broad philosophical approaches to estimating uncertainty using statistics: the frequentist framework (the basis for most standard statistics) and the

Box 4.4. Communicating uncertainty

In the key statements of high-level summaries, it is helpful to use agreed and calibrated language to express your level of certainty and uncertainty.

For quantitative analyses that lend themselves to formal statistical treatment, or for judgments where the experts are comfortable assigning broad probability ranges, the following reserved language can be used:

Virtually certain	Greater than 99% chance of being true or occurring
Very likely	90–99% chance of being true or occurring
Likely	66–90% chance of being true or occurring
Medium likelihood	33–66% chance
Very unlikely	1–33% change of occurring or being true
Exceptionally unlikely	Less than 1% chance of occurring or being true

For more qualitative statements, this language could be used:

		Amount of evidence		
		Limited	Medium	High
Level of agreement	High	Agreed but unproven	Agreed but incompletely documented	Well-established
	Medium	Tentatively agreed by most	Provisionally agreed by most	Generally accepted
	Low	Suggested but unproven	Speculative	Alternate explanations

Source: Moss and Schneider 2000.

Bayesian framework. In the frequentist framework, uncertainties are derived by hypothetical repetitions of the data collection process (i.e., taking multiple independent samples). In the Bayesian framework, uncertainties are derived from the laws of probability. In many cases the two approaches will produce the same estimate, and the choice of technique is mainly an issue of practicality or preference.

One advantage of the Bayesian framework is that it provides a means of combining different datasets or sources of information. This can be done in two ways: through use of a "prior," whereby the Bayesian model updates existing information in the light of new information, or through hierarchical models, whereby different datasets or information sources form their own "submodels," which are then combined at higher levels of the hierarchy to obtain estimates based on all the datasets. Meta-analysis is one such hierarchical approach and has been very useful in combining information from a large number of studies of tropical deforestation in order to determine the importance of different drivers of deforestation (Geist and Lambin, 2001).

Whether using a frequentist or Bayesian approach, it is important to keep in mind that all measures of uncertainty tend to underestimate the true uncertainty, since by definition only the "known unknowns" can be quantified—not the "unknown unknowns."

Assessments are often undertaken because contradictory or conflicting information exists. The assessment may not be able to resolve all of these. Some may be a result of legitimately different interpretations of the same data. Others may be because the data are inadequate for a definitive test of one theory over another. In these cases the task of the assessment is to be fair to all valid viewpoints, and it is the job of the review editors to ensure that this is achieved. Being fair does not necessarily mean equal space, and there is no need to provide a platform for discredited ideas. Assessments should not only point out disagreements. They should note where there is consensus or near consensus as well.

4.6.3 Presenting complex information

The body of the assessment is invariably words. The use of jargon and technical language should be minimized, and where it is unavoidable it should be supported by a glossary. Sentences that are short and simple work best, as do paragraphs that are short and on a single topic. In summaries for policymakers, paragraphs can start with the conclusion and then proceed to build the case for the lead statement. Varying the presentation of information can keep it interesting—for instance, by putting a table or picture on each double-page spread. Bullet lists can create punchy summaries of items that need not fall in a particular order, and numbered lists can be used for items in order of priority. In both cases, lists of between three and seven items are the most reader-friendly. Boxes or sidebars can provide information that is relevant but include a level of detail inappropriate for the main flow of the text or a slight digression. Appendixes are used for the same purpose, but more so—that is, for material that is relevant but bulky and not essential for a top-level reading of the text.

Tables give the reader the actual data. They are often there for reference purposes (so that someone can look up an exact value for a place or time). It is very hard to comprehend tables with more than six elements in their totality. Readers can be helped by providing row or column totals, highlighting certain cells, and providing guidance in the table caption about what they should especially notice. Vertical and horizontal lines in tables should be used sparingly and judiciously, in order to make the table uncluttered and easy to read. All columns need to be headed with a description and the units of measure. Superscripts in the table can be linked to footnotes to supply essential detail on data sources, exceptions, or comments.

Graphics are used to show relationships and trends in the data. Scientists tend to follow certain conventions in graph layout, in terms of how the axes are labeled and in keeping the graphic uncluttered. Graphs with fewer than six lines have been found to work best. Continuous variables should be presented as line graphs rather than bar graphs. The standard deviation (or confidence interval) should appear on the graph. Fancy effects (such as 3-D graphs, picture backgrounds, etc.) are best avoided unless they are essential to the message. Captions for graphs and tables need a generous amount of information. Many people read assessments by just looking at the pictures and reading the captions, so it important to make sure they can stand alone. It is not necessary to repeat in the text the information in a caption, and in general it is not advisable to show exactly the same information both as a table and as a graphic.

Maps and satellite- or model-derived images form an important part of many assessments. Although this makes them expensive to print, color is often indispensable for these forms of communication. Even if it is not essential, color printing is often worth the investment to make an attractive report. (The same applies to using good-quality paper and the skills of a professional layout person.) The use of maps and images in communication materials is an art form in itself, but some broad guidelines are available. The amount of detail in an image needs careful thought. This relates to how many categories there are in the legend, the colors or patterns chosen to represent them, and how large the map or image will be in the final report. It helps to reduce clutter—but without deleting essential orienting information such as latitude and longitude grid marks, scale bars, north arrows, and place names. Map or image legends should be sufficiently detailed that the user does not have to search in the text for basic explanatory detail. And the images should not be so small that they cannot be read.

Photographs and line drawings (e.g., cartoons or sketches) can often reinforce a message very effectively and make the overall text much more readable and appealing. Photographs that include people need the informed consent of those in the image.

Many decision makers find the forest of technical graphs and tables unintelligible; they often consider word-based narratives (anecdotes or stories) and case studies much easier to understand. It is important, therefore, to cater for both styles of information acquisition. Some kinds of information, such as describing the pathway between ecosystem services and human well-being impacts, lend themselves far better to a narrative style or diagrams than to quantification.

References

Alkire, S. 2002. Dimensions of human development. *World Development* 30 (2): 181–205.

Balmford, A., L. Bennun, B. ten Brink, D. Cooper, I. M. Côté, P. Crane, A. Dobson, N. Dudley, I. Dutton, R. E. Green, R. D. Gregory, J. Harrison, E. T. Kennedy, C. Kremen, N. Leader-Williams, T. E. Lovejoy, G. Mace, R. May, P. Mayaux, P. Morling, J. Phillips, K. Redford, T. H. Ricketts, J. P. Rodríguez, M. Sanjayan, P. J. Schei, A. S. van Jaarsveld, and B. A. Walther. 2005. The Convention on Biological Diversity's 2010 Target. *Science* 307:212–213.

Biggs, R., R. J. Scholes, B. J. E. Ten Brink, and D. Vackár. 2007. Biodiversity indicators. In *Sustainability indicators: A scientific assessment*, ed. T. Hák, B. Moldan, and A. L. Dahl. Washington, DC: Island Press.

Carpenter, S. R., W. A. Brock, and D. Ludwig. 2007 Appropriate discounting leads to forward-looking ecosystem management. *Ecological Research* 22 (1): 10–11.

CBD (Convention on Biological Diversity). 2003. *Monitoring and indicators: designing national-level monitoring programmes and indicators*. Montreal: CBD Secretariat.

CBD (Convention on Biological Diversity). 2007. *An exploration of tools and methods for valuation of biodiversity and biodiversity resources and functions*. Montreal: CBD Secretariat.

Cook, R. B., R. J. Olson, P. Kanciruk, and L. A. Hook. 2001. Best practices for preparing ecological datasets to share and archive. *Bulletin of the Ecological Society of America* 82 (2): 138–41.

Costanza, R., R. d'Arge, R. de Groot, S. Farber, M. Grasso, B. Hannon, S. Naeem, K. Limburg, J. Paruelo, R.V. O'Neill, R. Raskin, P. Sutton, and M. van den Belt. 1997. The value of the world's ecosystem services and natural capital. *Nature* 387:253–260.

Dasgupta, P., and K-G. Maler. 2001. *Wealth as a criterion for sustainable development*. Henley-on-Thames, U.K.: NTC Economic & Financial Publishing.

Dasgupta, P. 2007. The socio-economics of science and sustainable development. Plenary Lecture prepared for the International Conference on Science and Technology, Tokyo, September 7–8.

DEFRA (Department of Environment, Food and Rural Affairs). 2007. *An introductory guide to valuing ecosystem services*. London: DEFRA Publications.

Geist, H. J., and E. F. Lambin. 2001. *What drives tropical deforestation? A meta-analysis of proximate and underlying causes of deforestation based on subnational case study evidence*. University of Louvain, Belgium: LUCC International Project Office.

Gelman, A., and J. Hill. 2007. *Data analysis using regression and multilevel/hierarchical models*. New York: Cambridge University Press.

Howarth, H. B., and Matthew A. Wilsdon. 2006. Theoretical approach to deliberative valuation: Aggregation by mutual consent. *Land Economics* 82 (1): 1–16.

IUCN (International Union for Conservation of Nature). 2002. *2002 IUCN red list of threatened species*. Gland, Switzerland: IUCN.

IUCN (International Union for Conservation of Nature) and UNEP (U.N. Environment Programme). 2003. *World database on protected areas*. Gland, Switzerland, and Nairobi: IUCN and UNEP.

Kinzig A., C. Perrings and R.J. Scholes. 2009. Ecosystem Services and the Economics of Biodiversity Conservation, ecoSERVICES Group Working Paper, Phoenix.

Loh, J., ed. 2002. *Living planet report 2002*. Gland, Switzerland: World Wide Fund for Nature.

MA (Millennium Ecosystem Assessment). 2003. *Ecosystems and human well-being: A framework for assessment*. Washington DC: Island Press.

Magurran, A. E. 2004. *Measuring biodiversity*. Oxford: Blackwell.

Moellering, H., ed. 2005. *World spatial metadata standards: Scientific and technical characteristics, and full descriptions with crosstable*. Amsterdam: Elsevier, for International Cartographic Association.

Moss, R.H. and S.H. Schneider, 2000. Uncertainties in the IPCC TAR: Recommendations to lead authors for more consistent assessment and reporting. In: *Guidance Papers on the Cross-Cutting Issues of the Third Assessment Report of the IPCC*, R. Pachauri, T. Taniguchi, and K. Tanaka (eds.), World Meteorological Organization, Geneva, 33–51.

Myers, N., R. A. Mittermeier, C. G. Mittermeier, G. A. B. da Fonseca, and J. Kent. 2000. Biodiversity hotspots for conservation priorities. *Nature* 403:853–58.

Prescott-Allen, R. 2001. *The wellbeing of nations: A country-by-country index of quality of life and the environment*. Washington, DC: Island Press.

Rees, W. E., and M. Wackernagel. 1994 Ecological footprints and appropriated carrying capacity: Measuring the natural capital requirements of the human economy. In *Investing in natural capital: The ecological economics approach to sustainability*, ed. A. Jansson, M. Hammer, C. Folke, and R. Costanza. Washington, DC: Island Press.

Sanderson, E.W., M. Jaiteh, M. A. Levy, K. H. Redford, A. V. Wannebo, and G. Woolmer. 2002. The human footprint and the last of the wild. *BioScience* 52 (10): 891–904.

Schneider, S.H., B.L. Turner, and H. Morehouse Garriga, 1998. Imaginable surprise in global change science. *Journal of Risk Research*, 1(2), 165–185.

Scholes, R. J. 2009. "Ecosystem services: Issues of scale and tradeoffs." In *The Princeton guide to ecology*. S. R. Carpenter, H. Charles, J. Godfray, A. P. Kinzig, M. L., J. B. Losos, B. Walker & D.S. Wilcove. Eds. Princeton, NJ: Princeton University Press.

Scholes, R. J., and R. Biggs (eds.). 2004. *Ecosystem services in southern Africa: A regional assessment*. Pretoria, South Africa: CSOR.

Scholes, R. J., and R. Biggs. 2005. A biodiversity intactness index. *Nature* 434:45–49.

European Communities. 2008. *TEEB: The economics of ecosystems and biodiversity*. Brussels: European Communities.

Ten Brink, B. J. E. 2000. *Biodiversity indicators for the OECD environmental outlook and strategy*. Bilthoven, Netherlands: National Institute for Public Health and the Environment.

Tomich, T.P., C.A. Palm, S.J. Verlarde, H. Geist, A.N. Gillison, L. Lebel, M. Locatelli, W. Mala, M. van Noordwijk, K. Sebastian, D. Timmer, and D. White. 2005. *Forest and Agroecosystem Tradeoffs in the Humid Tropics. A Crosscutting Assessment by the Alternatives to Slash-and-Burn Consortium conducted as a sub-global component of the Millennium Ecosystem Assessment.* Alternatives to Slash-and-Burn Programme, Nairobi, Kenya

UNEP (U.N. Environment Programme). 2002. *Global environmental outlook 3*. London: Earthscan.

UNGA (United Nations General Assembly). 2001. *Road map towards the implementation of the United Nations Millennium Declaration*, Doc A/56/326, New York, 6 September.

5

Scenario Development and Analysis for Forward-looking Ecosystem Assessments

Thomas Henrichs, Monika Zurek, Bas Eickhout, Kasper Kok, Ciara Raudsepp-Hearne, Teresa Ribeiro, Detlef van Vuuren, and Axel Volkery

What is this chapter about?

This chapter offers guidance on how to set up a scenario exercise and how to develop, analyze, and use scenarios within ecosystem assessments. It sets out to provide a detailed overview of all the important steps that need to be kept in mind when embarking on a scenario exercise, providing in-depth analysis and indispensable background material on all key decisions that need to be taken.

Section 5.1 introduces scenario development and analysis as an approach to exploring uncertain and complex future developments in a structured manner. Also, it reflects on how a scenario exercise may support an ecosystem assessment. Section 5.2 summarizes practical considerations in setting up scenario exercises. Although such exercises are ideally tailor made to fit their context, an exemplary approach to developing scenarios based on the so-called deductive method is outlined in section 5.3. The implications of assumptions made in scenarios can be analyzed either by qualitative or quantitative means; section 5.4 introduces different approaches to analyzing the implications within individual scenarios as well as comparing outcomes of assumptions about the future across sets of ecosystem scenarios. Finally, section 5.5 reflects on the use of scenarios for scientific exploration and research, for education and public information, and for decision support and strategic planning.

The information provided in this chapter offers a starting point to planning a scenario exercise in the context of an ecosystem assessment. However, we stress that each scenario process is different and ideally requires a tailor-made approach. Thus the chapter does not provide a universal step-by-step approach nor does it attempt to provide a set of authoritative guidelines to scenario development.

5.1 How to explore the future with a scenario exercise

Section's take-home messages

- In addition to analyzing current ecosystem state and trends, forward-looking assessments need to explore the prospects of future developments: scenario exercises provide a structured approach to addressing related uncertainties and complexity.

- Scenarios can be defined as plausible and often simplified descriptions of how the future may unfold based on a coherent and internally consistent set of assumptions about key driving forces, their relationships, and their implications for ecosystems.
- Scenario exercises can serve different purposes: to support scientific exploration and research, to inform education and collaborative learning processes, or to underpin decision processes and strategic planning.

The state and functioning of ecosystems, and thus their ability to provide services, are subject to constant change resulting from the complex interplay of various driving forces. Ecosystem assessments need to not only provide a picture of the current state but also look into "the future" to assess the effectiveness of options for addressing environmental change (WRI 2008). However, the aspiration to understand "future" changes requires assessing developments that—while they may have their origins in past or current trends—have not happened yet (and may or may not happen). As a result, assessments of the future of ecosystems need to deal with considerable degrees of both complexity and uncertainty.

The complex nature of ecosystems results from the various interactions that govern ecosystem dynamics as well as from the multiple anthropogenic stressors that have an impact on the environment or lead to changes in the provision of ecological services. Such stressors (or direct drivers) include pollution, climate change, hydrological change, resource extraction, and land degradation and conversion. In turn, these direct drivers result from long causal chains of indirect socioeconomic drivers, such as demographic, economic, and technological developments. Finally, changing patterns of human values, culture, interest, and power set the conditioning framework (or ultimate drivers) for unfolding socioecological systems (MA 2005a; see also Chapters 3 and 4). This complexity makes it important for any forward-looking ecosystem assessment to capture as many of these interactions as possible by using a systemic framework that includes key economic, social, and environmental subsystems and the links between these (see Chapter 3). Another important feature of complex systems is that changes do not necessarily occur gradually or linearly but can be abrupt or accelerate once certain critical thresholds are crossed. Indeed, crossing a threshold can have substantial—often irreversible—impacts on ecological and social systems (WRI 2008).

Two principal types of uncertainty further complicate assessments of future environmental change. The first uncertainty arises from an incomplete understanding of the interactions and dynamics within ecosystems. Through recent assessments scientists have been able to considerably enhance their understanding of processes within ecosystems, but it is still important to recognize that there are elements that cannot fully be explained, and it is likely that there are dynamics within ecosystems that the scientific community is unaware of to date. (See Chapters 3 and 4 for a more detailed discussion of this first type of uncertainty.) As a result of this ignorance, any assessment of future changes starts off from an incomplete understanding of the current situation.

A second type of uncertainty is the indeterminacy of all future developments. Three distinct sources of such indeterminacy are ignorance, surprise, and volition (Raskin et al. 2002, MA 2005a). Ignorance here refers to limits in scientific knowledge in the understanding of possible future dynamics. This can be—but need not be—a result of complexity, and it is similar to the first type of uncertainty. Surprise is uncertainty due to the inherent unpredictability of complex systems that can exhibit emergent phenomena and structural shifts (for example, due to their underlying, determining socioeconomic and ecosystems dynamics). Volition is the unique uncertainty that is introduced when human actors are internal to a system under study and the future is subject to human choices that have not yet been made (MA 2005a).

Scenario exercises are seen as being particularly useful for assessing the prospects of future developments within complex and uncertain systems, such as ecosystems. In the Millennium Ecosystem Assessment (MA), scenarios are defined as "plausible and often simplified descriptions of how the future may develop based on a coherent and internally consistent set of assumptions about key driving forces and relationships" (MA 2005a). This definition captures the key features of most scenarios, although alternative definitions of what exactly constitutes a scenario have also been put forward (see, for example, IPCC 2000, UNEP 2002, EEA 2005). Nearly all definitions have in common that scenarios explore a range of plausible future changes—and they usually stress that scenarios are neither predictions nor forecasts or attempts to show the most likely estimates of future trends (see Figure 5.1).

While the remainder of this chapter focuses on describing scenario exercises as a tool, we do note that a host of other approaches can be used—and often have been—to address the prospect of "future" changes in ecosystems assessment. Some of these methods can be used instead of scenario-based approaches, while others are valuable alongside or as part of a scenario exercise. Such methods to assess future trends include trend analysis and trend extrapolation, forecasting (e.g., Armstrong 2001), cross-impact analysis (e.g., Gordon and Hayward 1968), future workshops (e.g., Jungk and Müllert 1987), Delphi-type expert-based estimates (e.g., Helmer 1983), role playing, gaming and simulation, and future state visioning (e.g., Stewart 1993), as well as developing future histories, science-fiction writing, or even wild speculation.

Within the field of scenarios, different types can be distinguished. Several typologies have been offered to characterize scenarios according to their key characteristics. Commonly, scenarios can be grouped according to their principal format, the main issue or type of question addressed, the process applied to develop the scenario, or the epistemology underlying a scenario exercise (see, for example, Box 5.1; see also Ducot and Lubben 1980, Ogilvy and Schwartz 1998, EEA 2001, van Notten et al. 2003, Börjeson et al. 2006, Westhoek et al. 2006, Alcamo and Henrichs 2008, Wilkinson and Eidinow 2008).

Maybe the most straight-forward distinction between scenario types relates to the format of a set of scenarios—that is, the differentiation between qualitative and quantitative scenarios. Qualitative scenarios are predominantly presented as narrative descriptions of future developments, commonly in the form of phrases, storylines, or images. Quantitative scenarios, in contrast, expand on numerical estimates of future developments—presented as tables, graphs, and maps—and are often based on the output of simulation modeling tools (Alcamo and Henrichs 2008). While this distinction appears clear cut in theory, many scenarios published to date are hybrids of these two types: selected aspects of qualitative scenarios may

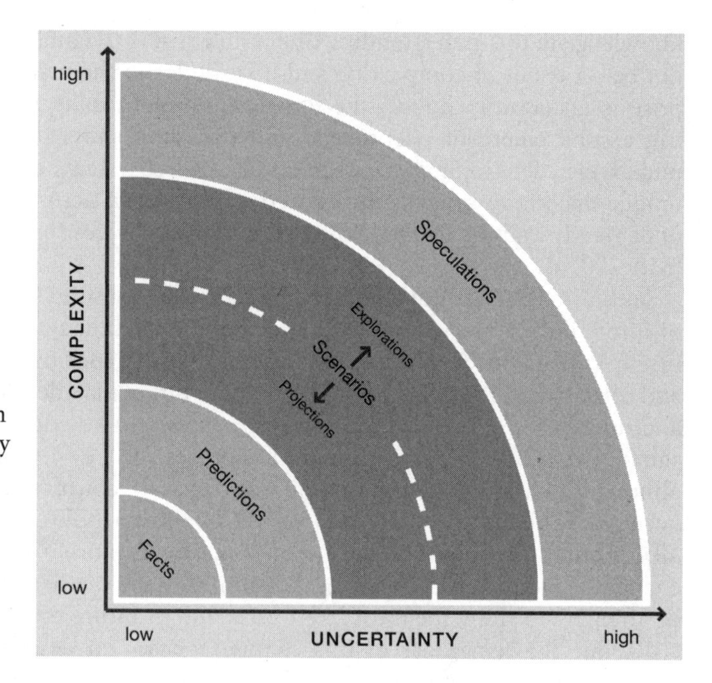

Figure 5.1. Scenarios can help address uncertainty in complex systems; note that scenarios differ from facts, forecasts, predictions, and speculations. *Source:* Zurek and Henrichs 2007

be underpinned by numerical estimates, or the numerical estimates of a quantitative scenario may be bound together and explained by a consistent storyline. Indeed, many of the recent international environmental assessments have developed and analyzed scenarios that explicitly combine qualitative and quantitative information (see EEA 2001).

An important function of scenario analysis—particularly in the context of ecosystem assessments—is that it provides an approach to reflect on and think through the possible implications of alternative decisions in a structured manner. Simply put, a scenario exercise offers a platform that allows decision units (individuals, a company, an organization, or even a country) to reflect on how changes in their respective context (that is, developments not within their immediate spheres of influence) may affect their decisions. This approach of testing whether different policy and management approaches are robust is sometimes referred to as "wind tunneling." However, for scenarios to be useful to "test" or "wind-tunnel" decisions in such a manner, the individual unit's actions and the decision context need to be distinguished clearly—which is all the more difficult to do in complex systems, as numerous feedbacks between the decision unit's behavior and contextual developments exist. Methodologically, this provides a significant challenge, and many past scenario exercises in an environmental or ecosystem context do not separate these spheres clearly, making it difficult to use the scenarios developed by them in concrete decision-making situations (see section 5.5).

Scenario-based approaches have evolved to be a useful and much applied approach to support environmental assessments over the past few decades (see Box 5.2). Also, ecosystem assessments have increasingly made use of scenarios to explore the potential future implications of different approaches for sustaining ecosystem services in the face of growing pressures. The MA, for example, offers four global

Box 5.1. Examples of scenario typologies

A first example distinguishes scenarios according to the type of question about future developments that a scenario exercise sets out to address. Three principal types can be differentiated by these criteria.

- *Reference scenarios.* Sometimes also referred to as "predictive scenarios." Generally set out to address the question "what is expected to happen?" and include forecasts as well as what-if analyses.
- *Explorative scenarios.* Attempt to map "what can or might happen?" and explore what future developments may be triggered either by exogenous driving forces (developments that are external and cannot be influenced by the decision makers in question), by endogenous driving forces (developments that are internal and can be influenced by decision makers), or by both.
- *Normative scenarios.* Sometimes referred to as "anticipatory scenarios." Aim to illustrate "how can a specific target be reached?" or "how might a specific threat be avoided?" and thus include both backcasting studies and planning exercises. Börjeson et al. (2006) offers a more detailed discussion of these types of scenarios.

A second example groups scenarios based on the epistemologies that underpin their exercises. Again, three principal types can be differentiated.

- *Problem-focused scenario exercises* center on the factors shaping future developments and usually emphasize the product rather than the process.
- *Actor-centric exercises* focus on the relationship of specific actors to their environment and primarily see scenarios as a basis for strategic conversations (particularly in an organizational learning context).
- *Reflexive interventionist scenario processes* are developed around the interactions between various actors and their environment (and vice versa) with the aim to inform action learning (especially in a public policy context). Wilkinson and Eidinow (2008) offer a more detailed discussion of these types of scenarios.

scenarios based on the implications of different assumptions regarding approaches toward governance and economic development (regionalized versus globalized) and toward ecosystem service management (reactive versus proactive), as illustrated in Figure 5.2 (see MA 2005a, Carpenter et al. 2006).

Besides global-level scenario based assessment, many regional or sub-global ecosystems assessments—including several of those associated with the MA—have either used the MA scenarios to help frame regional assessments or have developed new regionally specific scenarios to inform their analyses (see Table 5.1; see also MA 2005b, Pereira et al. 2005, Kok et al. 2007).

5.1.3 Why, when, and for whom to develop scenarios within ecosystem assessments

There can be several reasons to look in the future in a structured manner. Ultimately, it can be argued, the aim is to attempt to anticipate possible consequences of current

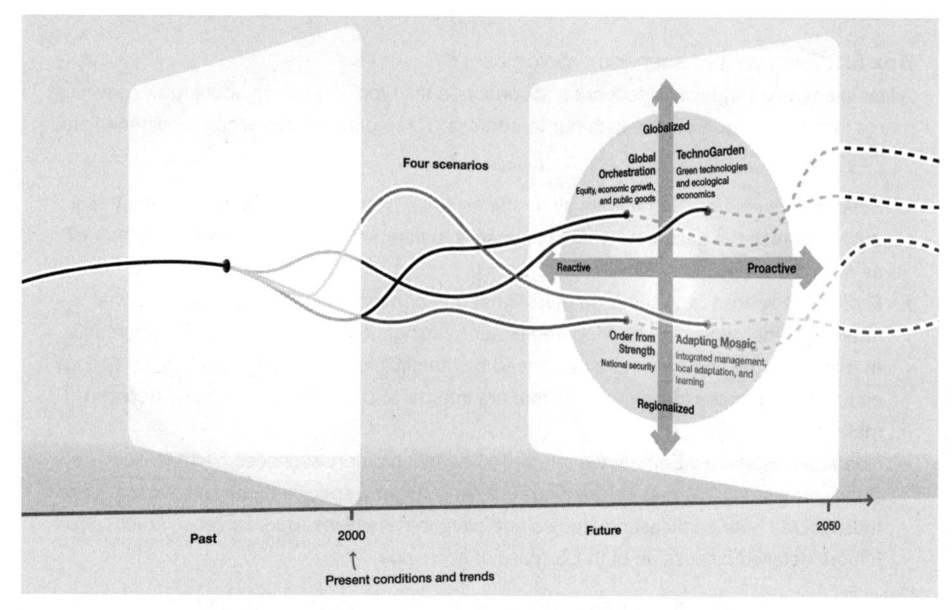

Figure 5.2. Millennium Ecosystem Assessment scenarios: plausible future developments through 2050.
Sources: MA 2005a

developments and options to either prevent, counter, prepare for, enhance, or benefit from future changes—and to better understand the implications of the uncertainties that surround assumptions about how the future may unfold. Supporting this overall aim, future studies in general—and scenario exercises in particular—can be used for multiple purposes (see Jaeger et al. 2007), including:

- Aid in recognition of "weak signals" of change;
- Avoid being caught off guard—"live the future in advance";
- Challenge "mental maps";
- Raise awareness (e.g., about future risks or critical thresholds);
- Test strategies for robustness using "what if" questions;
- Provide a common language (e.g., by unveiling different perceptions and beliefs);
- Stimulate discussion and creative thinking;
- Provide better policy or decision support; and
- Stimulate engagement in the process of change.

These and the other purposes underlying scenario exercises may be roughly grouped into three overarching clusters (based on, e.g., Alcamo and Henrichs 2008).

Scenarios developed and analyzed in support of research and scientific exploration can help to better understand the dynamics of (complex) systems by exploring the interactions and linkages between key driving forces. This cluster includes the group of "scientific scenarios" that sets out to examine the possible long-term behavior of biophysical systems as perturbed by human influence (MA 2005a). Also, it can offer a platform to bring together information from different research strands

Box 5.2. History of scenario development and analysis in the context of ecosystem assessments

The formalized use of scenarios as an approach to deal with uncertainty about future developments goes back more than 50 years. While humans have always made implicit and explicit "what-if" type assessments to guide their decisions, the earliest cited scenario studies are game analyses and military planning exercises published during the cold war (see, for example, Kahn 1960, Kahn and Wiener 1967). In the 1970s and 1980s the concept of scenario analysis was further developed and used for strategic planning within businesses (see, for example, Wack 1985). At the same time, the first scenario studies that explored environmental concerns emerged, often based on mathematical simulation models (see, for example, Meadows et al. 1972, Mesarovic and Pestel 1974).

Scenario analysis gained increasing prominence in the context of environmental assessments during the late 1980s and the 1990s—largely prompted by concerns with climate change and sustainable development (Raskin 2005). Various quantitative scenarios were used extensively to explore the consequences of technological and economic assumptions for energy use, greenhouse gas emissions, and climate change (see, for example, IPCC 1992, Alcamo et al. 1998). At the same time, more-qualitative scenarios were used to sketch out alternative environmental pathways (see WBCSD 1997). Also, the first examples of combined qualitative and quantitative scenarios were elaborated (see, for example, Toth et al. 1989, Gallopin et al. 1997, Raskin et al. 1998).

More recently, many—if not most—major global environmental assessment exercises have included a scenario-based component addressing future challenges (including, for example, Gallopin and Rijsberman 2000, IPCC 2000, IPCC 2001, UNEP 2002, MA 2005a, IPCC 2007, UNEP 2007; for an overview see EEA 2001 and Rothman 2008). It is worth noting that in recent assessments there has been some convergence toward a set of "stereotypical" scenario logics—which share the perspective on key uncertainties and assumptions about different driving forces (see, for example, van Vuuren et al. 2008):

- *Economic optimism scenarios*, which have a strong focus on market dynamics and economic optimism usually associated with rapid technology development (e.g., the *A1* (IPCC) or *Markets First* (UNEP) scenarios).
- *Reformed market scenarios*, which also focus on market dynamics but include some additional policy assumptions aimed at correcting market failures with respect to social development, poverty alleviation, or the environment (e.g., the *Global Orchestration* (MA) or *Policy First* (UNEP) scenarios).
- *Regional competition scenarios*, which assume that regions will focus more on their more immediate interests and regional identity, often assumed to result in rising tensions among regions and/or cultures (e.g., the *A2* (IPCC), *Security First* (UNEP), or *Order from Strength* (MA) scenarios).
- *Global sustainable development scenarios*, which see a strong orientation toward environmental protection and reducing inequality, based on global cooperation, lifestyle change, and efficient technologies (e.g., the *B1* (IPCC), *Sustainability First* (UNEP), or *Technogarden* (MA) scenarios).
- *Regional sustainable development scenarios*, which focus on finding regional solutions for current environmental and social problems, usually combining drastic lifestyle changes with decentralization of governance (e.g., the *B2* (IPCC) or *Adapting Mosaic* (MA) scenarios).

Table 5.1. Summary of the scenario exercises in selected subglobal assessments under the Millennium Ecosystem Assessment framework

Subglobal assessment	Stated goals of exercise (primary purpose of scenarios, see section 5.2.2)	Main methods used
San Pedro de Atacama	communication with stakeholders (i.e., education & information)	workshops and expert work
Caribbean Sea	stimulate thinking about the future (i.e., education & information)	workshops and expert work
Coastal BC	n.a.	workshops and modeling
India Local	assess influences of external forces on local community (i.e., education & information // decision support)	based on "what . . . if" questions for management options
PNG	change ways of thinking about the future (i.e., education & information)	assessment and implications of the past; expert scenarios
Portugal	for users and decision makers (i.e., scientific exploration // education & information)	workshops and expert work
SAfMA	tool for planning/actions particularly at local scales (i.e., education & information // decision support)	participatory workshops including community theatre (at local level); modeling and expert work (at basin and regional level)
Sweden KW and SU	prepare for surprises, information for planning; obtain stakeholder input (i.e., scientific exploration // education & information)	expert work
Northern Range	stimulate thinking about the future (i.e., education & information)	focus groups for developing storylines
Tropical Forest margins	analyze natural resource management options, future planning; enhance participation; inform policy makers (i.e., education & information)	expert work (in Mae Chaem); participatory scenarios (elsewhere)
Downstream Mekong	visualize the future, information for policy makers, input for models (i.e., education & information // scientific exploration)	n.a.
Western China	information for the government, input for models (i.e., scientific exploration)	quantitative modeling
Sinai	for local communication (i.e., education & information)	workshops—qualitative
Bajo Chirripo	get in touch with user needs (i.e., education & information)	workshops—qualitative
Eastern Himalayas	improve response options; inform policy makers (i.e., education & information // scientific exploration)	workshops—qualitative and quantitative

Table 5.1. continued

Subglobal assessment	Stated goals of exercise (primary purpose of scenarios, see section 5.2.2)	Main methods used
São Paulo	envision the future; change bad situation (i.e., education & information)	assessment and implication of the pilot expert scenario
India Urban	share information with partners (i.e., education & information)	individual consultations; literature review
Wisconsin	improve ecological management (i.e., scientific exploration // decision support)	initial expert assessment and scenario development; participatory scenarios workshop; scenario redrafted by experts

Information in the original table was based on information from specially designed questionnaires.

Source: Based on MA 2005b.

and scientific disciplines to better examine the complexity of ecosystems or to explore whether systems are likely to cross critical thresholds beyond which changes are irreversible—and the implications of doing so. When the principal purpose of a set of scenarios is to support scientific research, it is important to ensure that the procedure for building them unequivocally conforms to good scientific practice and that the assumptions behind scenarios are scientifically plausible (Alcamo and Henrichs 2008). A large number of examples of this type of scenario exercises can be found in the scientific literature. Prominent international examples include the scenarios elaborated by the IPCC (IPCC 2000, IPCC 2007), the Millennium Ecosystem Assessment (MA 2005a), and the Global Environmental Outlook (UNEP 2002, UNEP 2007). (See also section 5.5.2.)

In the wider context of education and public information, scenarios can provide an approach for structuring, conveying, and illustrating differing perceptions about unfolding and future developments. At the same time, scenarios can help to highlight and explain the implications and long-term consequences of current trends and choices that may lie ahead. This cluster combines two groups of scenarios suggested in MA (2005a)—"new conversation scenarios" (those aimed at exploring new and unknown topics that be used as an educational tool for wide audiences, de facto offering a tool for collaborative learning exercises) and "groups-in-conflict scenarios" (those used to help understand differences in worldview and perceptions of groups and to jointly explore consequences of actions). If the main goal of building scenarios is to inform the general public or a particular target group, then it is particularly important that the scenarios are perceived to be credible, stimulating, thought provoking, and—most important—relevant to the audience (Alcamo and Henrichs 2008). (See also section 5.5.3.)

Scenarios also have a long history in decision support and strategic planning (see Leemhuis 1985, Ringland 2002)—to solicit views and opinions about expected future developments, to "test" different options for decision units to respond effectively to changing decision contexts, to evaluate the implications of specific decisions, to help prepare for risks and trends, or to analyze the trade-offs related to specific future pathways. Especially in the context of business and private enterprises, "business strategy scenarios" have been used to explore uncertainty in a decision context that an individual decision unit does not control, in order to test the robustness of options and to identify opportunities and challenges (MA 2005a). Many of these scenarios remain internal to the organization that develops them (as, especially in the realm of business, scenarios may give a competitive advantage; see Wack 1985, Schwartz 1991). By contrast, "public interest scenarios" or "strategic conversation scenarios" are commonly necessarily open to wider debate. These aim at shaping the future by articulating a common agenda and language between actors and highlighting potential actions and their consequences (MA 2005a) and by "unearthing" the assumptions about future developments that often guide decision making implicitly. In the field of public policy, Ringland (2002) shows a number of examples of scenario exercises that were directed to support policies at local scales. Furthermore, it is worth noting that in the context of environmental change, scenarios originating from a research context have played a major role in supporting decision-making processes (see also section 5.5.4).

Ecosystem assessments generally set out to be "a social process to bring the findings of science to bear on the needs of decision makers" (MA 2005a)—in other words, they aim to contribute to varying degrees to all three clusters just described (see also Box 5.3). As noted in Chapter 1, assessments have been seen to have the most impact when they are perceived to combine three characteristics: saliency to potential users ("Is it relevant?"), credibility with regard to use of scientific methods ("Is it sound and convincing?"), and legitimacy in the way the exercise is designed and conducted ("Is it inclusive and unbiased?"). (See also Mitchell et al. 2006.) Alcamo and Henrichs (2008) argue that this is true also for scenario exercises, particularly those developed in the context of an assessment process.

Ideally, any scenario exercise would thus aim to be relevant, credible, and legitimate at the same time. Furthermore, scenarios can benefit from aiming to be creative and to challenge prevailing expectations and worldviews (see Alcamo and Henrichs 2008). Practically, however, time and resource limitations often require setting priorities in meeting one or two of these characteristics rather than all of them. But also conceptually these notions may be somewhat mutually exclusive within scenario exercises: in many circumstances it may be difficult to be fully inclusive and still be convincing (for example, when different scenarios cater to contrasting worldviews and beliefs). Or when a scenario exercise is designed to explore the implications of possible future surprising events, this may entail assessing nonlinearities that cannot be constructed using scientific methods alone. Thus while each of the above characteristics is worthy of pursuit in its own right, there may be particular need to emphasize one of them depending on the main purpose of the scenario exercise—and to be aware of what trade-offs between these concepts are acceptable. (This is discussed in more detail in section 5.5.)

5.2 How to set up a scenario exercise

Section's take-home messages

- A scenario exercise can have both process- and product-related outcomes: typically such exercises lead to an intense learning experience for those involved as well as a set of scenarios for further analysis and use.
- It is crucial to clearly define the purpose and goals of a scenario exercise at its outset, as this will affect the set up, scope, and planning of the exercise—it will also guide the scenario development process and the usefulness of the outcomes.
- Key aspects to consider when setting up a scenario exercise are—among others—its context, timing, budget, degree of stakeholder participation and expert involvement, desired complexity, and geographical scale of analysis.
- Scenario exercises are complex processes and require both an enabling authorizing environment and a sound organizational set up: The value of involving experienced scenario practitioners when designing the process cannot be emphasized enough.

There is no "one-size-fits-all" approach to conducting a scenario exercise. How a scenario exercise is set up and carried out very much depends on its context, goals, participants, and so on. Nevertheless there are a number of stages and steps common to many, if not most, exercises. The following sections describe some of the key steps and offer a basic approach based on ecosystem assessment experiences with scenarios. However, the steps of the modular process outlined here can and should be adapted to fit specific needs. The way to carry out a scenario exercise depends a great deal on its context; and to avoid being bogged down by the confusion and preconceptions that often surround discussions about the future, it is thus crucial to be clear about the goals and the set up of the exercise.

Generally speaking, careful planning in the early stages of a scenario exercise significantly improves the quality of any scenario development process and its outcomes. Putting together a plan at the outset will help guide the development process through the identification of the specific goals, steps, and resources that are necessary for a meaningful analysis. Nevertheless, any plan should offer enough flexibility for revisiting and adjusting any of the steps outlined here, as much will be learned throughout the process, and the approach decided on in the initial phase should not be seen as static (Jaeger et al. 2007).

Before embarking on a scenario exercise to assess the future of ecosystem services and human well-being, a few practical issues should be considered, which are explored in the next five sections:

- What to expect from a scenario exercise—possible outcomes;
- How to frame the purpose of the scenario exercise and decide on the scenario type;
- How to define the scope of a scenario exercise;
- How to establish an authorizing environment and project team for a scenario exercise; and
- How to facilitate participation throughout a scenario exercise.

Box 5.3. Reflections on the role of scenario exercises in ecosystem assessments

The eventual focus of a scenario exercise very much depends on when in an assessment process it is performed. It is useful to distinguish four principal settings. The four roles described here are not mutually exclusive—but clarity about the primary aim of a scenario exercise within an ecosystem assessment is essential for ensuring that scenarios are developed, analyzed, and used in the best possible fashion.

First, a scenario exercise at the outset of an ecosystem assessment may be a useful process to help participants develop a common understanding and framework about those dynamics that shape future ecosystem developments. In particular, it may aid in arriving at a common language about current trends, threats, opportunities, and options. Such a scenario process should ideally involve key experts and preferably also important stakeholders in the outcome of an assessment. However, it may not be necessary to embark on a full scenario exercise, as the key aim of this exercise is to arrive at a shared way of thinking about future changes.

Second, a scenario exercise may be run in parallel with other activities within an assessment—with the aim of having the different components of an assessment inform each other. In the Millennium Ecosystem Assessment, the Scenarios Working Group worked at the same time as the Current Status and Trends and the Policy Responses Working Groups, so that ideas developed in the Current Status and Trends Group helped shape the discussion of drivers in the Scenarios Working Group, and vice versa. However, running a scenario exercise in parallel to other activities may make it difficult to incorporate key findings across different Working Groups unless sufficient time is set aside for iterations between them to allow fully exploring and harvesting respective findings.

Third, if an assessment is structured in a more linear fashion, a scenario exercise can provide a link between a discussion on current trends and the potential of policy responses to counter undesirable developments. In such cases, a scenario exercise would build fully on the outcomes of an analysis of the current state and trends within an ecosystem and adapt—to the extent possible—the framework, structure, and language developed. Similarly, the viability and robustness of policy responses would be tested against the backdrop of the different scenarios developed. This would greatly help achieve overall consistency across different parts of an assessment; it also entails the risk of limiting discussions.

Fourth, a scenario exercise may be limited to supporting outreach activities only, using the outcomes and findings of an assessment to develop scenarios at various scales with stakeholders and decision makers, molded around their immediate concerns. As an example of such a setup, an assessment would describe current trends and outline possible policy responses—and based on the wealth of information developed there, different future scenarios might be explored. In this case, the discussion of scenarios describing future developments is not a part of the actual assessment; rather, the assessment is the input in one or more follow-up scenario exercises.

5.2.1 Setting up a scenario exercise—what outcomes to expect

Generally speaking, scenario exercises have two main types of outcomes: a *process* (i.e., a learning experience for actors involved in the scenario development) and a *product* (i.e., the actual set of scenarios themselves, which can be used in different ways).

Outcome 1: The process—benefits of the scenario development process

Scenario exercises can have three primary process-related benefits. First, those who participate in a scenario development process gain a better understanding of interactions, assumptions, and trade-offs related to ecosystem services and human well-being. Some of the direct outcomes of scenario exercises are an increased understanding of the linkages between the different parts of the socioecological system, the identification of beliefs and assumptions about how a policy or a chosen development pathway may alter some or all of the system, the identification of potential long-term consequences for ecosystem services of choices made in the near future, and the identification of factors important for a successful outcome of a decision.

Second, scenario exercises create a platform to talk across interest groups, disciplines, and philosophies. Uncertainty about the future has an equalizing effect: no one discipline or sector can predict the future. Scenario work requires scientists, governments, and citizens to collaborate and piece together plausible stories about what might occur in the future. The result is a process that can accommodate thoughtful, creative, and nonthreatening discussion about topics that might otherwise be politically charged. Less powerful groups can be empowered through such a process, and more powerful groups can gain invaluable insight into how their practices and policies affect others. Providing space for multiple forms of knowledge (e.g., traditional and practitioner) can lead to deeper and more nuanced reflections on social and ecological change within a system.

Finally, the discussion of and reflection on different scenarios can create the grounds to reveal conflicts, exchange information, and help to build consensus or at least an understanding over controversial issues related to ecosystem services and the choice of polices for sustaining services. Indeed, scenario exercises can be used to air conflicts or build consensus among diverse stakeholders over what a desirable future might look like. Managing natural resources often involves trade-offs between different values and economic activities. Getting stakeholders around the same table to discuss their visions of future land management or economic development helps build understanding of these trade-offs and agreement on appropriate policy. When developing and discussing a set of scenarios jointly, hidden values and assumptions are uncovered, highlighting potential shared values and the root of conflicts. Taking stakeholders away from the present day to focus on possible futures facilitates discussion, allowing participants to develop a greater understanding of each other's point of view. While there is no guarantee that increased mutual respect will carry over to resolving current conflicts, it increases that possibility.

Outcome 2: The product—key elements of scenarios

Most scenarios, including those developed in an environmental and/or ecosystem context, tend to have a number of key elements in common: a representation of the initial situation, key driving forces, a description of step-wise changes, and image(s) of the future (see, for example, EEA 2001 or Alcamo and Henrichs 2008).

The first key element of any scenario is an understanding of and representation of the initial situation of the system, including an understanding of how past trends have shaped the current state. This information may be gathered or developed as input into a scenario exercise or as part of it. The description of the current state of the system can stem from other parts of an ecosystem assessment. In the Millennium Ecosystem Assessment, for example, information from the Condition and Trends Working Group, which described the prevailing conditions and trends of ecosystem services and human well-being and their relation to past and current drivers of change, was used to inform the scenarios developed.

Driving forces are the main factors that influence future developments and dynamics of the system a scenario focuses on. Main categories of driving forces in scenario exercises focusing on ecosystem services include the so-called STEEP drivers (social–cultural, technological, economic, environmental, and political driving forces). It has proved useful to distinguish between direct driving forces, also known as structural drivers (those that unequivocally influence the system) and indirect driving forces (those that alter the level or rate of change of one or more direct drivers). Another practical distinction is between endogenous drivers (which are in the control of the decision makers in a given system) and exogenous ones (which are outside of the decision makers' control). (Also see Chapter 3 for more information.)

A further important element of any scenario is a plausible and consistent description of step-wise changes that are assumed to unfold in the future. These changes are based on assumptions about how key driving forces develop and interplay, and how this affects the state of a system at different points in time. These changes can be depicted as numbers and figures (see section 5.3) or as sets of phrases, illustrative vignettes, and/or detailed storylines. The number of time steps described within a given period may vary according to the focus of the scenarios developed and the need for analytical underpinning (see Alcamo and Henrichs 2008).

Finally, one of the end products of a scenario exercise is a description of an image or several images of the future, describing in great detail what the future may look like as a consequence of the assumptions made about drivers and their step-wise changes. Often, the system's "end state" (its state at the time horizon of the scenario) is presented in the form of a narrative description (which may be more or less extensive, depending on the objectives of the exercise). Indeed, illustrated and narrative descriptions of the future have often proved useful for communicating the outcome of scenario assumptions to a wider audience (see section 5.5).

5.2.2 How to frame the purpose of a scenario exercise

Scenario exercises are much more likely to be useful if their purpose is clearly identified and spelled out right from the beginning. The purpose will guide how the process should be organized, who should be involved in the exercise, and what the scenario should focus on. Because it is possible that many of the audiences and participants in a scenario exercise will be unfamiliar with such efforts, it is important to have clarity about the main goals and outcomes in order to communicate effectively with potential participants and end users, achieve good buy in, and encourage effective participation. More important, clarity about the goals of a scenario exercise will help to determine what type of scenario is needed, to what degree the scenario will need to be analyzed using simulation tools, to what level an exercise needs to be geared toward strategic conversation rather than detailed analysis, to what extent a set of scenarios should aim at covering expected trends versus being mind-stretching exercises, and so on. The purpose(s) of the exercise may evolve or be expanded, but a clear starting point is necessary.

Thus, at the outset those in charge of putting together the scenario exercise will need to discuss and decide on specific goals of the exercise. As part of this process, stakeholders and decision makers might be interviewed to help focus the goals further. Through examination of the goals of the assessment and consideration of the resources available (including time, funds, and expertise) as well as the needs and interests of the different stakeholders, the assessment team and principal stakeholder groups can decide which general type of scenario exercise they will develop.

Questions that may be useful to determine the goals of a scenario exercise include:

- Why is the scenario exercise being initiated?
- What stakeholders are most interested in the scenarios component of the assessment and what kind of information are they interested in?
- What should be gained from developing scenarios in terms of concrete actions?
- What policies/plans/projects will be informed by the scenario exercise?
- What type of scenario (for example, qualitative or quantitative scenarios) is required to produce information that is useful to the principal stakeholders?
- What limitations are relevant to this exercise in terms of time, expertise, and logistics?

The answer to these and similar questions very much depends on the general context of the exercise—that is, whether it is geared toward scientific exploration, education and information, or decision support. The general context can guide what type of scenario exercise might usefully be developed, and along what parameters. The following paragraphs present some ideas on how the scenario exercise might be developed depending on which context it fits into. In practice, scenario exercises will often have several goals and therefore may fit more than one context, and the process should be adjusted accordingly.

In a scenario exercise that aims to support research and scientific exploration of future links between ecosystem service trends and human well-being, the focus is more strongly on the scenarios themselves rather than the process. The participation of people from a range of scientific disciplines—and to some extent, from stakeholder groups—is key to gaining new insights into the links between ecosystem services and human well-being. In order to complement the assessment of current and past conditions and trends, it may be important for the scenario process to be coordinated time-wise with the rest of the assessment so that accurate information about ecosystem service trends can be used as a starting point from which to develop scenarios. (This is also true when the goal of the exercise is to inform decision making.) Several brainstorming sessions with a large scenarios team can help promote the creative development of storylines. The scenario team can try to incorporate more complexity into the scenarios and examine cross-scale interactions. Particular attention should be paid to thresholds, risks, probabilities, and the possibility of surprise in the system of interest. End results may include qualitative storylines and quantitative models; the identification of key scientific uncertainties, risks for human well-being, and the sustained flow of ecosystem services; and future research needs. (See also section 5.5.2.)

In a scenario exercise that aims primarily to provide education or (public) information, broad participation is crucial. A range of stakeholders and decision makers needs to be involved, and the role of scientist and researchers may be limited to providing input rather than being involved in the actual scenario development process. Discussions might be aimed at challenging mental models, creating buy in for the assessment, or communicating with key decision makers in a nonthreatening setting. Stakeholder workshops can be used to discuss the system of study and possible future trajectories in a way that is engaging and emphasizes thinking "outside the box." Such an exercise can also be empowering for marginalized stakeholder groups through knowledge sharing and building adaptive capacity for dealing with ecosystem change. Scenario sessions can be run strategically by a neutral facilitator to ensure a tight focus, clear understanding on the part of the participants, and as much active participation as possible. Inspiring texts, theatre, art, and public forums can be used as alternative formats to spread the results to a larger audience. (See also section 5.5.3.)

Finally, in scenario exercises geared toward decision support and strategic planning, a greater degree of caution is required in order to develop scenarios that are aligned with the problems of interest to decision makers and at the same time based on a process that is perceived to be legitimate. (Key questions here are "who has developed the scenarios?" and "with what agenda did they do so?") Policy or normative scenarios may support decision making by providing an analytical framework for finding the most suitable policy options with regard to a specific policy target. (See also Box 5.4.) Exploratory scenarios may aim at providing a backdrop for "strategic conversations" that challenge and sharpen the mental models of decision makers (in this, the purpose does not differ vastly from the goals of education and information). Either way, in the context of decision making it is crucial to engage in a constant dialogue with those who take the decisions. It is worth noting that the effectiveness of scenarios in this context is particularly dependant on their acceptability to the dominant decision culture. Ideally it is the decision makers and stakeholders themselves who drive the scenario development process, with scientists acting as "resource people." However, in some situations—in particular at the global scale—this is not possible. Separate scenario sessions with different groups of experts and decision makers may be a strategic, alternative way to develop plausible and useful scenarios and ensure cross-fertilization without overly taxing the time of decision makers. It is important to iterate several times between groups in order to provide the specific information desired by decision makers. (See also section 5.5.4.)

After discussing the general goals of the scenario exercise, an assessment team can put together a list of process parameters and objectives that seem to be important for achieving the goals of the exercise, without getting into too much detail

Box 5.4. Scenarios in the policy cycle

In the case of scenario exercises geared toward supporting policy-making processes, the goals and design will depend on when in the policy process the scenarios are developed. The role of scenarios varies according to the different phases of a policy cycle (see figure) and may include identifying early warning signals and evaluating evidence during the phase of problem definition; illustrating potential consequences if a problem remains untreated, which can be useful in the phase of agenda setting; or evaluating the feasibility of policy options that are compatible with the phase of policy formulation.

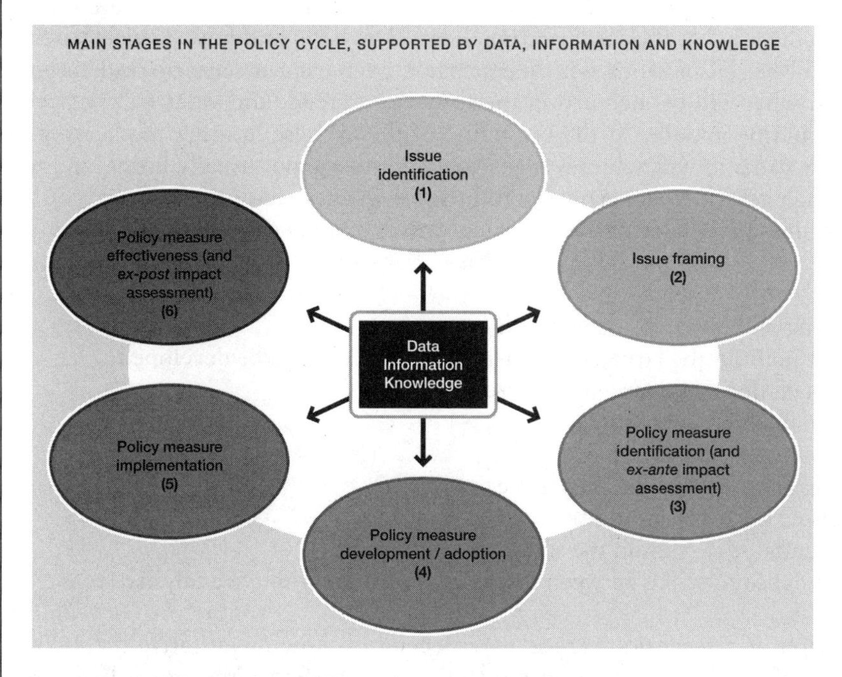

Figure. The policy cycle is a "heuristic" framework that breaks down the policy-making process into several phases (Source: EEA 2006)

(for example, determining whether an exercise should be more or less quantitative, should involve high-level decision makers, and so on). There are no rules about how all the components fit together; instead, members of the scenario team should familiarize themselves with different options and decide how to best meet their goals.

An additional important consideration that might affect the deliberations relates to the question of what comes after the scenario exercise. All too often this is only considered after a set of scenarios has already been developed and analyzed. However, this question is more usefully posed at the start of the scenario planning phase in order to determine why and how scenarios will be used and therefore how they might be developed effectively. Sometimes the purpose of the exercise is to come to a better understanding of the problem under investigation, and no follow-up action is needed. The same could apply if the core objective was to bring together a group of stakeholders and initiate a common learning process. However, more often than not a scenario exercise is not an end in itself but also a means to support other activities and processes, and the eventual scenarios will be used in the context they were designed for (and sometimes beyond). (See section 5.5.)

5.2.3 How to define the scope of a scenario exercise

Once the general purpose and the specific goals of a scenario exercise have been established, it is useful to develop an overview plan. It is often unclear to assessment teams that have not undertaken scenario work previously how much of an investment of time and resources is necessary to complete a full exercise. This usually depends on how complex the scenarios are to be, whether there will be a modeling element to them, what the scenario time horizon will be, and how participative the exercise will be. Discussions with participants over what a scenario and scenario exercise is, what will be included in the scenario exercise, and what will not be included should be initiated at the beginning of the exercise in order to clarify what the exercise can and will achieve. Scenario exercises are not usually linear, and they require much iteration between steps to produce the desired outcomes.

Determining the scope of the exercise involves making decisions about a number of issues:

- How long will it take to develop the scenarios?
- At what point in the larger assessment will the scenarios be developed?
- How much does a full scenario exercise cost?
- How to involve stakeholders?
- How to involve experts?
- Which issues to focus the scenario exercise on?
- What time horizons should the scenarios address?
- What spatial scale should the scenarios address?
- What balance to strike between qualitative and quantitative analysis?

The duration of a scenario exercise may depend on how quantitative the analysis will be and if participants rely on modeling or workshop discussions. If the time available to complete an exercise is extremely short, planners might consider using and/or adapting existing scenarios from other exercises (as described later in this chapter). In general, several months are required for preparation and development of a scenario exercise at the sub-global scale. If many stakeholders are involved, much more time may be needed just to get everyone together. In the MA, the exercises ranged from one participatory workshop that lasted a weekend (but required several weeks of preparation and follow-up work) to a series of short workshops that spanned several months.

The timing of a scenario exercise largely depends on the information it depends on and how outcomes will feed back into the rest of the assessment. While scenarios can be useful at any point in the larger assessment process, most sub-global assessment teams in the MA initiated scenario exercises toward the end of their larger integrated assessment process. This had the advantage of having the assessment of conditions and trends of ecosystem services provide a starting point for the scenarios. However, many of these exercises subsequently found that it would have been preferable to start them a bit earlier in order to inform the main assessment at the same time as being informed by it. The benefits of holding a scenario exercise early in the process include jump-starting a positive participatory process around a creative and nonthreatening activity, the early identification of stakeholder concerns and priorities, and the identification of uncertainties that require further assessment by the core assessment team. One sub-global assessment under the MA process (in Wisconsin, in the United States), for example, began the integrated assessment with a scenario exercise. This allowed them to identify key uncertainties and dynamics in their system with the participation of stakeholders, and these findings could then be used to focus the larger assessment (Peterson et al. 2003).

The available budget will often dictate whether meetings and workshops can be held and therefore how much stakeholder participation is possible. Scenario workshops commonly require a minimum of two sessions of one or two days each, but several additional workshops would be better and allow for more in-depth discussion. Local participants may be willing to set aside a day at no cost (other than meals and logistics), but a regional or global exercise may need to fund travel and accommodation. Funding may also be needed to cover the time of experts involved in analyzing the scenarios. Particularly in model-based scenarios, analysis can become resource intensive, as described later. Budgets will thus vary widely depending on the context of the exercise.

The level of stakeholder involvement may have an impact on the timing of the exercise and the complexity of the scenarios. How many stakeholders to involve, and in what capacity, strongly depends on the goal of the scenario exercise. This is discussed in more detail later in this section.

The level of expert involvement depends on what kind of scientific and local expertise is available to help develop the scenarios. Generally speaking, expert involvement may strengthen an exercise's credibility—but it does not guarantee it. Experts may be involved as core members of the scenario team, or they may be able to offer feedback on the developed scenarios. Another important consideration in this context is whether experts need to be paid for their involvement—again this depends largely on the exercise context.

The degree of complexity of the scenarios depends on how many themes and/or ecosystem services and/or drivers of change and/or indicators will be explored. Particular ecosystem services and drivers of change will already have been identified as especially relevant to an assessment context during the assessment of conditions and trends in ecosystem services and human well-being. (See Chapter 4.) This identification process is also part of developing a conceptual framework that contains the main factors and services and their relationships. The conceptual framework can then be used as a tool to guide the scenario development as it provides a unifying way of thinking and conceptualizing many relevant issues.

The time horizon of a scenario should be based on what is a reasonable amount of time for the main issues of concern to be explored or managed. "Time horizon" here refers to the period of time over which the scenarios will be allowed to unfold. While 30 years might be an appropriate timeframe for scenarios developed to explore land use and land planning, more time might be needed to examine climate change impacts and policies. Time horizons also have political implications and cannot always be selected in advance of the initial exploration of issues with stakeholders about policy cycles and information needs. In the Millennium Ecosystem

Assessment, for example, scenarios were explored against the time horizon of 2050.

The geographical scale for a scenario exercise is not always easy to determine, particularly if complex driving forces interact across geographical or organizational scales and levels. (See Box 5.5 for additional guidance on the issue of scales.) In most cases the geographical scale of the assessment (e.g., global or local) also determines the scale for the scenarios. Nevertheless, scenarios developed at a specific scale can also aim to address multiscale issues and relationships—especially given that recent work on socioecological systems stresses the need for understanding processes at multiple scales and in particular their interaction across scales. Further information on how to link scenarios across scales can be found in section 5.3.

The primary mode of analysis—that is, quantitative or qualitative—will greatly depend on whether sufficient data and modeling tools are available, whether the budget allows for extensive quantitative analyses, and what kind of scenario outcome will be most suitable for the scenario exercise's purpose. In a scientific context, for example, scenarios may benefit from quantifying as many relevant parameters as possible. However, if no one is available to develop adequate models to support a scenario analysis, then a more qualitative approach will be the only option. In an educational or informational context, a scenario exercise's outcomes may be conveyed more easily using narratives and qualitative analysis, making detailed quantification superfluous or even counterproductive. Thus one approach toward answering this question for a specific exercise is to read published accounts of previous scenario exercises and compare the described process and outcomes with your own process and information needs.

There are additional questions that the assessment team may need to consider depending on the particularities of the scenario exercise. In some cases the team will wish to explore specific international environment or development targets; in other cases the focus will be on alternative development paths at local and regional scales, examining trends in a few important ecosystem services only. These ideas have to be reconciled with logistical limitations and the specific needs of important end users of the scenarios. The assessment team can hold a scoping meeting with key scenario end users to identify or refine a list of important ecosystem services and drivers of change and to decide on the most relevant spatial and temporal scales. The output would be a detailed overview plan of the scenario-building process that will serve as a flexible blueprint for the exercise.

5.2.4 How to establish an authorizing environment and project team for a scenario exercise

A scenario exercise's *authorizing environment* refers to the level of support offered for the scenario process and products by stakeholders and funding agencies. Setting up an authorizing environment for an ecosystem assessment in general is covered in Chapter 2, and many of the considerations highlighted there apply also to scenario exercises. Developing a strong authorizing environment revolves around building mechanisms to ensure the credibility, legitimacy, and relevance of the scenario exercise and its outcomes—which depends on how the exercise is run and how stakeholders are treated. It is therefore beneficial for a scenario exercise if stakeholders participate in or are consulted throughout the process to ensure a continued match

Box 5.5. Why worry about scale?

One issue that most scenario exercises in ecosystem assessments have to deal with is how to treat processes that play out at multiple scales. *Scale* refers here to the physical dimensions, in either space or time, of phenomena or observations, whereas *level* is used to describe the discrete levels of organization, such as individuals, households, ecosystems, or agroecological zone. Components of any complex system are structured hierarchically in space and time across scales and levels—and hierarchy theory suggests that the best way to deal with problems in a multiscale complex system is to understand how the elements of the system behave at a single time–space level.

Geographical scale simply refers to the boundary of the case study and—when models are used—the spatial resolution of the grid cells. These do not always overlap with the boundaries of the socioeconomic or organizational levels (i.e., the functional scale of relevant processes). Particularly in multiscale scenario exercises, selecting the appropriate temporal scale is very important. There is a general tendency for processes to become slower as the geographical scale expands. As a result, global assessments often consider a temporal extent of 2050 to 2100, especially when climate change is one of the processes of interest. In contrast, local studies often consider a time horizon of 10 years or even shorter, being in line with the policy cycle.

More often than not, within any case study important drivers and processes cover multiple temporal and geographical scales. It is therefore important to consider them accordingly when developing scenarios. The simplest way to consider multiple scales is by executing a single scale assessment and including the most important drivers from higher scales as external drivers. The essential difference to a multiscale assessment is that drivers at only one scale are dynamic. In other words, if the majority of all-important drivers are active at one scale—and exogenous drivers can be considered constant over space and time—a single scale assessment can be sufficient. Conversely, if the aim is to address drivers from different scales, while explicitly including changes in and between those drivers, a multiscale study is appropriate.

between their needs and the exercise outcomes. A few issues particular to setting up an authorizing environment for a scenario exercise—beyond those indicated in Chapter 2—also need to be taken into account.

How an appropriate authorizing environment is established will again very much depend on the context of a scenario exercise. If the context is research and scientific exploration, a structured and transparent scientific approach will be needed that includes identifying levels of uncertainty associated with future trends, based on what is known from the literature about ecosystem services and human well-being. For decision support, the relevant decision makers should be involved in the process from the beginning and recognize its legitimacy. In many cases, it is very useful to have stakeholders involved in describing the decision-making context central to the scenarios. For information, communication, and learning processes, a broad range of stakeholders may be included in the process and be given a chance to contribute meaningfully to either the storyline development or the interpretation of the scenarios. A lot of time usually needs to be allocated to discussions among participants of a scenario exercise (e.g., stakeholders and scientists) to promote learning and ownership of the process and outcomes.

With the assessment team and key stakeholders, it will be useful for the process and for the end users to put together a short document outlining what steps will be taken toward establishing an appropriate authorizing environment that will lead to a credible, legitimate, and relevant scenario process and outcomes. In setting up an authorizing environment for a scenario exercise, the following might be considered:

- Identify which organization or group of people is convening the scenario exercise. Are they trusted by the stakeholders who are involved? Is there a trusted, neutral facilitator running the exercise?
- Identify participants that represent the groups that the exercise is designed to reach. What will be their role in the exercise?
- Identify other people who may be affected by the changes being explored in the scenario exercise. What is their role in the exercise?
- Identify people who understand the system in question and who could help build credible scenarios. Are these people well respected by the relevant decision makers and scenario end users?
- Identify people from different disciplines and sectors who could review the scenarios and/or offer informal feedback on the process and outcomes.
- If possible, make sure decision makers are included in the actual exercise. If the decision makers do not have the time to participate in workshops, might they participate in another manner? Consider interviews, as well as regular briefings about the process.
- Are the scenarios being built around the expertise of the scientists involved or around the needs of the decision makers? This balance may need to be adjusted if the former is being promoted (unless this is the main focus of an exploratory exercise). Note that the needs of the decision makers may change as the process develops and more about the system is learned.

Planning, organizing, and facilitating a scenario exercise is a lot of work and requires a group effort, which commonly a core scenario team will be tasked with. Therefore, an important early step when preparing for an exercise is to establish such a team and assign responsibilities and tasks within it. The core team should include a coordinator who provides leadership and ensures the smooth running of the whole exercise. If the scenario exercise is part of a larger ecosystem assessment, it is advisable to include some members of the larger team alongside the experts and key stakeholders who will work only on the scenario. Building scenarios benefits from the inclusion of people representing different disciplines, different spheres of knowledge, and different societal roles, but the composition and balance of the team will depend on the goals of the exercise. As scenarios within ecosystem assessments will be broadly focused around the links between ecosystem services and human well-being, it is important to aim for a broad representation of scientific disciplines in the core team, including both natural and social sciences. The scenario exercise will be organized and driven by this core scenario team, often inviting a broader group of stakeholder representatives to participate in the scenario development process. The degree to which this latter group participates is discussed in section 5.2.5.

The scenario team's organizational set up can range from very inclusive to formal and strategic, and this too depends on the purpose of the exercise. If the goal of the exercise is outreach, a broad and inclusive scenario team may be desirable.

If the goal is scientific exploration, a smaller group of creative people from a range of scientific disciplines may be preferable. If the goal is decision support, again the group might be smaller, but it could include the key stakeholders who will inform the development of the scenarios according to their needs. Also, the focus of the scenario exercise will affect who might be on the scenario team. Scientists and practitioners who have a deep understanding of the different components of a system and how they are interconnected may effectively inform complex and nuanced narratives about possible futures of a region. Linked, multiscale scenarios addressing ecological dynamics may require the participation of scientists with specific technical skills. Assembling the scenario team as early as possible in the larger assessment process facilitates coordination of the process and information flows between the scenario exercise and the rest of the ecosystem assessment work (Evans et al. 2006).

Scenario exercises that rely heavily on stakeholder input or on intensive interactions between scientists and stakeholders benefit tremendously from including a facilitator as a member of the scenario team. The facilitator is tasked with moderating meetings and encouraging effective participation within the exercise. Generally, a facilitator should be someone who has run a scenario exercise before or who has taken the time to learn from other people's experiences and developed (and perhaps tested) a plan for the exercise. In some cases, it may be important for the facilitator to be able to communicate with people in their own language and style, while in other cases a facilitator from a more "neutral" background may be preferred. Being very familiar with the steps in a scenario exercise is of primary importance. Evans et al. (2006) provide many tips on facilitation within diverse groups of participants. Some key abilities in a facilitator include being able to paraphrase and summarize participants' inputs to clarify main points, encouraging participation through comments and body language, having strategies for dealing with difficult participants, being able to move the process along without sacrificing meaningful discussions, capitalizing on the diversity within a group by encouraging discussions about differences in perspectives and worldviews—and, most of all, being an unbiased moderator.

Other tasks within the core scenario team include:

- Keeping track of communication between the team and the stakeholder groups and making sure that the team is meeting the expectations of the end user;
- Planning logistics, running workshops, and putting together background reading materials;
- Taking notes at meetings, discussions, and workshops; and
- Developing illustrations, vignettes, and graphs to summarize important findings of the exercise for communication purposes.

5.2.5 How to manage participation throughout a scenario exercise

The many benefits of ensuring effective participation in scenario development processes are increasingly recognized, especially in scenario exercises geared toward supporting decision making or strategic communication. While scenario exercises have often been viewed as simply being a procedure for developing a product, they are now seen and used as a process that aims to involve the potential users of the scenarios (see section 5.5).

An important reason for involving stakeholders in scenario development is to enhance the legitimacy and the potential impact of scenarios. This can be a crucial factor in the usefulness of scenarios to support public decision making. At the same time, stakeholder participation can help tap into the expertise and creativity of stakeholders or experts. Involving stakeholders can guide emergent (social) learning processes within public, research, or policy communities (Alcamo and Henrichs 2008). Also, bringing together people working in different domains, many outside science, within a scenario exercise is being recognized as central to understanding the dynamics governing complex systems more completely. Scenario exercises with broad stakeholder involvement can thus function as a way of pulling together often very different knowledge sources and epistemologies.

A tested working model combines a core scenario team that organizes the exercise (usually composed of experts in the area of analysis and scenario practitioners) with a broader stakeholder group that contributes to development of the scenarios (this latter group is sometimes called a scenario panel). The broader stakeholder group is selected by identifying stakeholders who have an interest in the topic of the exercise and/or who could be affected by its outcome and then selecting a group representing the various stakeholders to participate in the exercise. Together, the core scenario team and the broader stakeholder group generally hold a lot of knowledge about the socioecological system, but individuals will vary in the information they have about the system and will have different agendas and biases that will inevitably lead to some conflict within the group. The core scenario team is tasked, therefore, with planning how they might incorporate the different forms of knowledge from the stakeholder groups in order to achieve a desired outcome from the process.

The rules of participation are best set up from the beginning. Talking openmindedly about the future is difficult for some groups, as people have different ideas about how much control they have over the future. Participants should therefore know exactly how and when they will be contributing and how much influence they will have over the storylines in comparison with the core scenario team. A talented facilitator is essential to managing this balance and retaining the interest and trust of all participants, as described earlier. Ensuring that the scenario process benefits from the input of a relevant cross-section of society increases the likelihood that the scenarios will have buy in from the appropriate actors, affecting the potential impact of the scenario exercise.

Also, the role of stakeholders in the scenario exercise should be discussed and clarified before involving stakeholders to avoid misunderstandings and risking the legitimacy of the process. Guidelines may be drawn up that should be shared with the stakeholders in order to promote transparency and trust. Participation in scenario exercises can be organized in different ways and may involve helping to define the system parameters of interest, participating in the development of scenarios, using the end products, or offering guidance on the process and contents of the exercise (Volkery et al. 2008). Simply put, stakeholders may either be consulted or can lead or co-lead a scenario development process (see Table 5.2).

Participation is both time consuming and sometimes expensive, and therefore a balance needs to be struck between having effective and useful participation and staying within the budget and time constraints of the exercise. If there is no budget

Table 5.2. Explaining different degrees of stakeholder involvement and the corresponding roles of stakeholders and core scenario team

Degree of stakeholder involvement	Stakeholders	Core scenario team	Explanation
Consultation	Supportive	Lead	The core scenario team develops the content and process with inputs from stakeholders via consultations and iterative reviews of the products
Co-design	Content supportive, process co-lead	Content lead, process co-lead	The core scenario team shares responsibility for the process with stakeholders, but leads the shaping of the scenario content
Co-decision	Co-lead on content and process	Co-lead on content and process	The core scenario team shares the development of the process and content with the stakeholders fully, often with the help of an outside facilitator
Decision	Lead	Supportive	The core scenario team supports the process of scenario development, but the stakeholders are completely responsible for both the process and content of the exercise; an outside facilitator may be used, or in some cases the core scenario team may facilitate

for hosting scenario workshops, less costly means of engaging stakeholders may be preferable. For example, several reviews of the scenarios by various stakeholders or consultations with key end users might be planned instead. Boxes 5.6 and 5.7 give examples of stakeholder participation approaches in two scenario exercises.

Again, the goals of the exercise guide how the different forms of knowledge might be incorporated and how stakeholders might contribute. In the context of developing scenarios for scientific exploration and research, the core scenario team will have a strong say in the storylines and scenario components to ensure credibility. Inviting experts from many disciplines to review the scenarios will help balance the inputs from participating disciplines and strengthen credibility further. An exercise aimed at scientific exploration might nevertheless try to encourage meaningful contributions from a range of stakeholders in order to capture new ideas, knowledge, and unknown dynamics within the system. Conversely, if the scenarios are developed to support decision processes, it is important to leave control of the storylines with the stakeholder panel in order to ensure legitimacy. The decision makers often have a broader understanding of at least the social components of the system and can help to design the system more realistically. An outreach exercise to communicate the scenarios to a broader audience will aim to have as open and inclusive a process as possible in order to achieve buy in from stakeholders and encourage broad collaborative learning.

Box 5.6. The PRELUDE scenario exercise

Over the course of a year, the European Environment Agency (EEA) organized three workshops to develop the PRELUDE scenarios, focused on the future of European land use and management (EEA 2007). Each workshop lasted three days. Experienced professional facilitators conducted the sessions with stakeholders to arrive at the final storylines. About 30 stakeholders were involved in the overall process, and although their travel and accommodation were paid for, they were not compensated for their time.

In the beginning, it was not clear whether it would be possible to maintain strong engagement with a large group of stakeholders with limited time in such an intensive process. The relative success of this endeavor can be explained by several factors:

- Full responsibility for drafting the storylines was given to the participants, and thus they developed an ownership of the process that made them return to subsequent meetings.
- The external facilitation of the process underlined that there was no hidden agenda of the sponsoring organization or any other involved organization. Observing this, the participants engaged in open and lively debate.
- Different animation exercises such as short movies were used to produce an inspiring working atmosphere that stimulated idea exchange and creativity among participants.

The scenarios developed by the stakeholders were strengthened and given credibility through reviews and quantitative modeling conducted by experts in relevant fields. The first workshop focused on identifying key uncertainties, driving forces, and the underlying scenario logics, as well as considering potential land use–related environmental impacts. After the workshop the draft scenarios were analyzed and reviewed by the EEA project team and a scenario analysis support group composed of land-use experts and modelers, who were also present at the workshop. The draft scenarios were then quantified using spatially explicit data from land use simulation models.

The objective of the second workshop was essentially to revise the first round of model results, check for inconsistencies, and refine the scenario storylines in view of the modeling data. Interaction between modelers and stakeholders resulted in the translation of the qualitative narratives into numerical trends. These numbers were further calibrated based on modeling data from existing relevant exercises.

The third and final workshop had three objectives: a final review of the five scenarios, a review of the environmental impacts of the scenarios, and a process to build consensus among participants concerning the final PRELUDE results, main products, and future dissemination activities. The outcome of the exercise is available at www.eea.europa.eu/prelude and is described in EEA 2007.

Balancing stakeholder involvement and expert knowledge within a time-bounded exercise is tricky. The more the stakeholders are encouraged to participate, the more complex and lengthy the process becomes, and time constraints must be managed strategically. In all cases, careful facilitation is important for managing power imbalances, language differences, and expectations among participants. No participants should ever feel that their contributions are being ignored.

Box 5.7. The Southern Africa Millennium Ecosystem Assessment (SAfMA) scenario exercise

The scenario exercises conducted within SAfMA took place at multiple scales. SAfMA-regional was an assessment of the African continent below the equator, SAfMA-Gariep was a water basin level assessment, and SAfMA-livelihoods was conducted within several villages in South Africa. At the regional and water basin scales, the scenario exercises were conducted entirely by the core assessment teams. SAfMA-regional synthesized the work of several sets of detailed scenarios that had been developed previously. The core team of SAfMA-Gariep developed original scenarios for the region. Both SAfMA-regional and SAfMA-Gariep then used their respective advisory boards to review the scenarios. The advisory boards were composed of representatives of key stakeholder groups in the region, and consultations with them were used to develop subsequent iterations on the scenarios.

At the local level, SAfMA-livelihoods invested in more time-intensive participatory approaches to developing scenarios. About 20 local participants were selected by their respective communities to be intensively involved in the scenario exercise based on the idea of so-called user groups. Each user group was defined by its main activity (e.g., livestock owners, fuelwood collectors, farmers, etc.), and members were selected by the community based on their participation in that activity, their willingness to participate, and their level of expertise. Each of the groups had to have a range of ages, from young people to elderly. Gender balance was achieved naturally because of the prevailing division of labor (i.e., fuelwood is usually collected by women; livestock is kept by men, etc.).

These user groups were involved over six months in exploring changes in the activities they were involved in, the resources related to these, and the drivers or underlying causes of those changes. These trends and observations were compared with the core assessment team's understanding of drivers that were taking place at broader scales during the same period. A group of actors used the whole set of information to make up storylines about the future based on this understanding of the past. The user groups then viewed these plays, changed them to better suit the reality in their community, and eventually the scenarios were performed for the entire community.

5.3 How to develop scenarios

Section's take-home messages

- Scenario development processes should be tailor made to meet a scenario exercise's specific needs. A selection of standard approaches exists, but these should always be adapted to the respective context and the scenario exercise's focal issue.
- Usually, scenario development consists of a number of stages that can be carried out in an iterative manner: identification of main concerns, discussion of key drivers and uncertainties, selection of underlying scenario logics, description of scenario assumptions, and analysis of the scenario implications.
- Scenarios can be depicted using either qualitative (such as stories or pictures) or quantitative (such as numbers and graphs) means—or combinations of these two. Translating qualitative descriptions into quantitative assumptions is not a trivial task, however.

5.3.1 Different approaches to developing scenarios

A host of different approaches can be and have been used to develop scenarios, very much depending on who builds the scenarios and for what purpose. The approaches to scenario development that might be used as part of an ecosystem assessment do not differ principally from scenario processes in other contexts—except that the primary focus here is on exploring how interactions between ecosystems and socioeconomic activities may unfold in the future. Generally speaking, approaches to scenario development can be roughly grouped into three types: deductive, inductive, and incremental. In this manual, we focus on introducing the key stages in a deductive type in order to illustrate the basic stages of a scenario exercise. For completeness, however, and to highlight key characteristics, we first introduce the inductive and incremental approaches.

Inductive approaches to scenario development require similar steps as a deductive approach but apply a different method for identifying key uncertainties and developing scenario logics. Rather than systematically discussing and deducing driving forces, this variant of scenario development starts off by describing individual events or plot elements and then spins larger stories around these seeds. By starting from episodic plot elements, developers can build a scenario that will have future consequences that call for some strategic decisions in the present. The process is somewhat less systematic than the deductive approach and calls for a degree of creativity and imagination that may be difficult to structure and coordinate, but it may result in more diverse and unusual scenarios (see Ogilvy and Schwartz 1998).

Incremental approaches build on expanding and questioning a "reference scenario" (or "business-as-usual scenario" or "official future"). A reference scenario is the future that scenario developers or decision makers really believe, either explicitly or implicitly, will occur. This is usually a plausible and relatively nonthreatening scenario, featuring no surprising changes to the current environment and continued stable growth. Then, to contrast this picture, alternatives are explored by first identifying key threats to this pathway and then varying the driving forces that appear most influential. Based on these, different interactions between key driving forces might produce unexpected outcomes and allow new scenarios to be built to contrast the original future outlook (see Ogilvy and Schwartz 1998).

Deductive approaches have four main stages: a first stage geared toward identifying main concerns about future developments; a second stage focusing on discussing the main uncertainties, driving forces, factors, and actors that might be expected to shape future trends and their interactions—and thus identifying the underlying scenario logics; a third stage during which the actual scenarios are developed; and a fourth stage in which the scenarios and their implications are analyzed. Each of these stages involves a number of steps and generates different types of information that can be relevant at different points in decision-making processes. Similar approaches to scenario development are also described by Jaeger et al. (2007) and Schwartz (1991).

As noted above, this manual focuses on introducing the key stages in a deductive scenario development. Figure 5.3 and Table 5.3 provide an overview of the basic stages and the information generated in each step:

- How to identify a focal issue (Stage 1)
- How to discuss drivers and scenario logics (Stage 2)

- How to describe scenario assumptions and scenario storylines (Stage 3)
- How to analyze scenarios—that is, the implications of scenario assumptions for the focal issue (Stage 4).

This approach allows a focus on exploring the most uncertain and most important driving forces in a systematic manner and is overall quite straightforward and easy to follow. However, note that while the approach is described here in a more or less linear way, there are likely to be iterations between the various steps. Furthermore, it is important to keep in mind that this approach to scenario development has the drawback of being rather rigorous and may not be the most appropriate in all circumstances—it very much depends on the purpose of the scenario exercise. Indeed, when embarking on a full-scale scenario exercise, the process should be tailored to the specific situation and context.

The remainder of this section and the next one focus on the outcomes that each scenario development stage needs to achieve in order to carry out a complete scenario exercise. However, they do not attempt to provide a universal step-by-step approach nor do they attempt to provide a set of authoritative guidelines to scenario development. For guidance on this, see Wollenberg et al. (2000), Evans et al. (2006), Jaeger et al. (2007), or Alcamo and Henrichs (2008). Also, this chapter does not aim to address in any detail the process and related (facilitation) techniques that might be used to arrive at the respective outcomes described in each stage—instead it stresses that involving trained facilitators to help guide the scenario development process is invaluable (see also section 5.2).

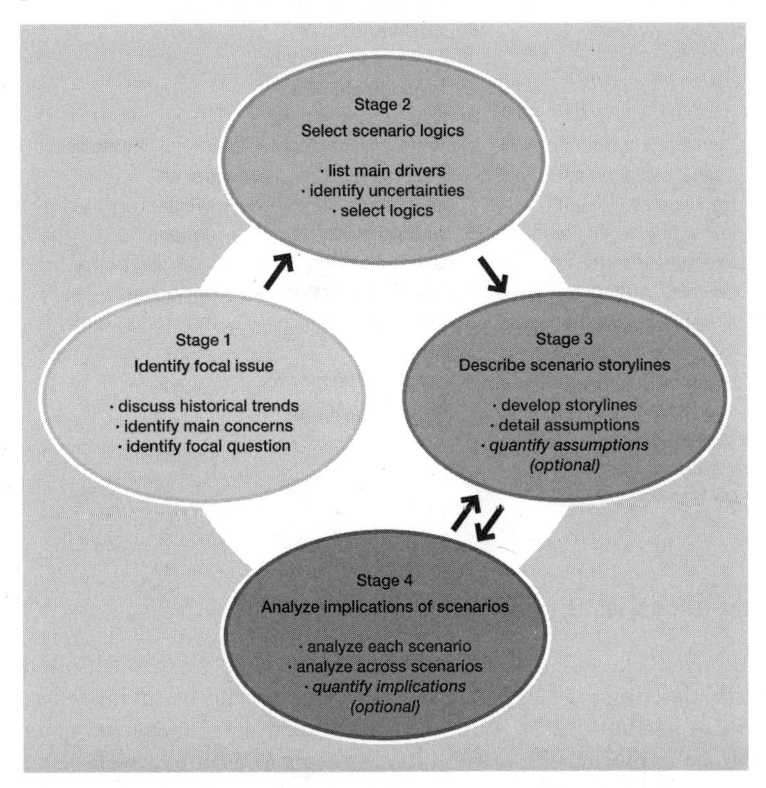

Figure 5.3. Key stages in a deductive-type scenario development process

Table 5.3. Steps in a deductive-type scenario development process and relevance to decision making

Scenario development stage	Steps *	Type of information generated	Relevance to decision-making processes**
Stage 1: How to identify focal issue	1) discuss historical areas and developments that led to present situation 2) identify main concerns for the future 3) identify focal questions (main problems) to be addressed by the scenarios	1) analysis of current problems and their roots, based on stakeholder analysis 2) analysis of key questions for the future 3) clear understanding of main assumptions for the future of the investigated system	– identifying issues – framing issues – identifying stakeholders to be engaged in decision process
Stage 2: How to discuss drivers and main uncertainties	4) list main drivers and uncertainties that will change the future 5) identify possible driver trajectories, thresholds, and uncertainty about them 6) identify main interactions between drivers	4) analysis of main drivers shaping the future and their importance 5) voicing of different view points on drivers' trajectories and their importance 6) understanding of system's interactions, development of a system perspective	– framing issues – prioritizing information – informing policy selection
Stage 3: How to describe scenario assumptions and scenario storylines	7) develop first drafts of scenario storylines 8) translate storylines into model inputs and execute a modeling exercise (optional) 9) finalize scenarios based on critical assessment of storylines (qualitative) and modeling (quantitative) results, as well as on stakeholder discussions	7) creative ideas about the future and emerging changes 8) challenges for assumptions on drivers' interactions, consistency checks 9) grounding of qualitative knowledge through modeling	– identifying decision points – evaluating policy options – selecting policy – designing monitoring systems

5.3.2 How to identify a focal issue (Stage 1)

When setting up a scenario exercise, the general scope will have been defined, often implicitly or explicitly determining the principal issues that should be addressed in a scenario exercise (see section 5.2.2). Within an ecosystem assessment the main issue would typically be exploring the future of ecosystems and human well-being in a particular region. However, the issue could also be analyzing the prospects for

Table 5.3. continued

Scenario development stage	Steps *	Type of information generated	Relevance to decision-making processes**
Stage 4: How to analyze scenarios	10) conduct analysis across the scenarios set 11) discuss scenarios analysis's results for various stakeholder groups 12) write up and disseminate scenario exercise	10) assessment of trade-offs and synergies of various management options 11) information to different stakeholders on differing view points 12) awareness of emerging issues for the future	– identifying policy options – evaluating policy options – developing strategies for policy implementation and monitoring

* Although the steps are described in a linear way, in practice there is much iteration among them.

** For more details on the policy cycle, see section 5.2 and Chapter 2.

Source: adapted from WRI 2008.

biodiversity and habitat loss, elaborating the trade-offs between different types of ecosystem services depending on different development paths, or similar issues.

It has proved to be helpful to clarify the questions that a set of scenarios should attempt to address further and to establish a so-called focal issue or focal question. For this, the main concerns and questions stakeholders might have concerning the future need to be identified. This can be done based on literature review, but it should ideally also involve a more iterative process, including interviews with stakeholders and decision makers. Based on the resulting understanding of the perceptions of the nature of prospective challenges, the impacts and dynamics of past developments (see below), and the prevalent expectations about future trends, a focal question for the scenario exercise can be constructed. It is desirable to make this focal question as objective and unambiguous as possible (Scearce et al. 2004) and preferably to link it to concrete choices, decisions, or strategic considerations at stake. Thus, rather than asking in general terms "What does the future hold for our ecosystem?" the focal question might be "Does the way we use ecosystem services in our region put future food security at risk?" or "Can the Millennium Development Goals be achieved, short term and long term?"

Typically, the answer to a focal question constructed in this manner will be "It depends!" One of the key aims of any scenario exercise is to distill what the answer depends on and what the related implications are. One of the main advantages of centering a scenario exercise on a concrete and unambiguous focal question is that it creates a common and transparent platform for the process. Particularly in the context of complex systems, it is easy for scenario developers to get lost in the myriad of interconnections and possible angles to an issue—a concrete focal question here helps to manage the degree of complexity somewhat (without neglecting the need to place the discussion around a specific focal question within its broader ecosystem and socioeconomic context). Consequently, focal questions guide the further scenario development and help identify which specific topics need to be included in the actual scenarios and which might be omitted—thus grounding the scenarios. Based on the focal question, developers will also have to discuss the time frame for the scenarios, balancing the short-term time horizon of many planning processes with the slow, long-term nature of ecosystem changes.

As noted, an early step in the process can include a look into the past to familiarize participants with changing development patterns and explore the wide range of factors that contributed to the current situation and problems. Identifying the main actors responsible for certain developments can be part of this analysis, which will later help to discuss the possible role of individual actors in the future and how the relationship among actors could change in different scenarios. Talking about past developments from different perspectives can also help create a common understanding of the present shared by all those who set out to construct the scenarios. Especially if the group developing the scenarios is composed of experts and stakeholders with varying expertise, this type of retrospective discussion is useful in order to establish a common starting point for the look into the future. Often it is illuminating to look as far back into the past as the group sets out to look into the future: This can help people realize how volatile systems can be and help scenario developers to get out of their current mind-set and take a more long-term perspective.

5.3.3 How to discuss drivers and select scenario logics (Stage 2)

The second stage extends the discussion to include an analysis of drivers of change (and if the scenario exercise is part of a broader ecosystem assessment, this discussion can build on the drivers already identified—see Chapter 4). Here the focus should be on the main drivers of change expected to play a major role in the future. For this, developers can distinguish between the drivers that directly affect the concerns identified in Stage 1 along with the underlying ones that seem more removed and indirect. Often it is the interaction of these indirect drivers that shape direct drivers' trajectories in the future, but they can be difficult to assess over shorter time frames. Exploring the web of drivers and how they are linked and influence each other is not an easy task. This discussion is likely to be shaped by different perspectives of scenario builders and differences in knowledge on each topic. Discussing the interactions between drivers can also help identify possible thresholds that will not just change the course of one driver but affect the functioning of the whole system.

The possible trends and trajectories for each driver should be discussed, such as the expected brackets of population numbers over the scenarios' time horizon or possible economic growth rates in the future. It is also important to identify how different participants perceive these trends and how certain they are they will play out in the future and why. This will help uncover participants' main assumptions about the future. The discussion of the certainty of how these trends will play out in the future is crucial, as one possibility is to develop the scenarios around the most uncertain drivers, exploring their possible different pathways in the future.

The discussion of direct and indirect driving forces can help identify the main uncertainties for the future, as any threshold points or bifurcation points identified can later help develop the actual scenario storylines. One important decision has to be made when working with this approach to identifying uncertainties: There can be uncertainty about a driver itself, its trend over the course of the future, or its interaction with other drivers that is the result of a lack of information (i.e., ignorance). In addition, there can also be uncertainty surrounding the impact that a driver and changes in its trend can have on the future. Clarifying which kind of uncertainty is more important for a particular exercise is helpful to avoid confusion among the scenario developers.

To prepare for a discussion of the underlying scenario logics and then build the actual scenario storylines, participants may find it helpful to rank all the drivers with respect to their importance for the identified problems and the degree of uncertainty (see Figure 5.4), document the reasons for the ranking, and discuss why and how the drivers interact. Reflecting on both types of uncertainty—what is known about a driver and about its impact—can be a useful tool for deciding on the critical uncertainties (see Jaeger et al. 2007 for additional information on how to identify critical uncertainties).

Based on this, the underlying scenario logics can be developed. One approach used in the past is to frame the scenarios based on two critical uncertain drivers that seem to influence most or all of the others (this has been used in a number of scenario exercises such as IPCC 2000 and MA 2005a; see Figure 5.5). Reaching agreement on these key uncertainties—usually the most important and most uncertain ones—is not an easy exercise, but once agreement is reached a matrix depicting the two opposing extremes of each uncertainty can be constructed. By combining the extreme assumptions of each of the two axes, four scenario logics emerge that can be used to develop the main stories of the scenarios.

It should be noted again that other approaches can be applied to establish scenario logics and that the "axes approach" is particular to the deductive approach

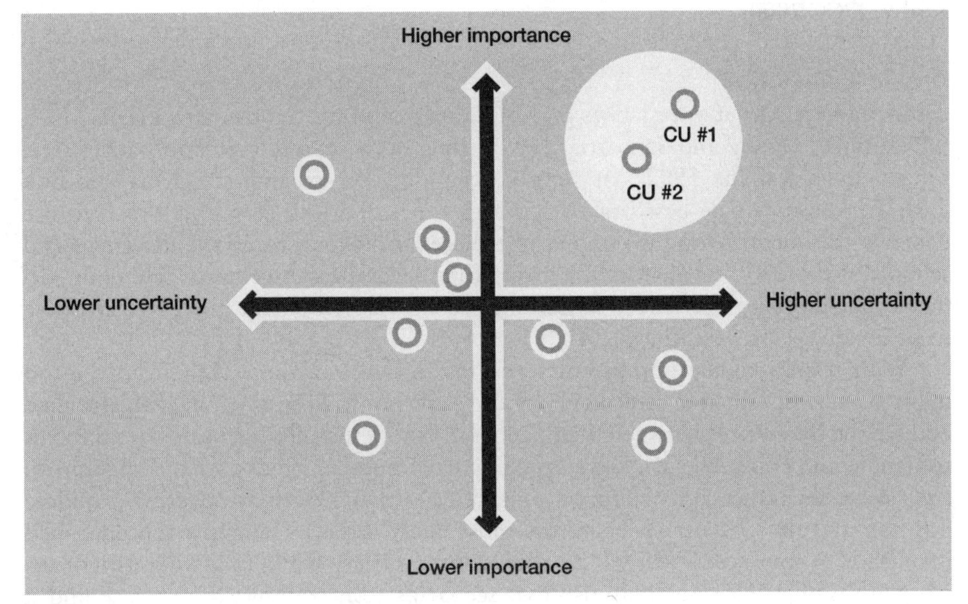

Figure 5.4. How to identify critical uncertainties.
Note: CU 1 and CU 2 denote critical uncertainties.
Source: Jaeger et al. 2007

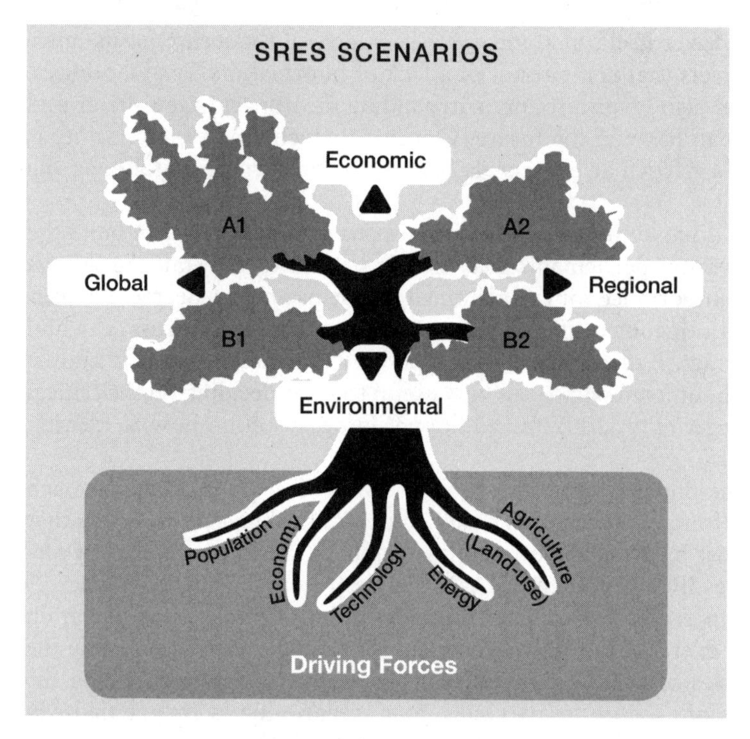

Figure 5.5. Schematic illustration of SRES scenarios. The four scenario "families" are illustrated, very simplistically, as branches of a two-dimensional tree. In reality, the four scenario families share a space of a much higher dimensionality, given the numerous assumptions needed to define any given scenario.
Source: IPCC 2000

described here. Other approaches may build on exploring "emblematic events" (see, for example, Ogilvy and Schwartz 1998), using decision tree type approaches (see, for example, Kahane 1992), or simply expanding on existing scenarios (see Box 5.8).There is no single best practice to arrive at scenario logics, and which to use depends considerably on the context of the scenario exercise and the dynamics and schools of thought prevalent in the group that develops the scenarios. Here, in particular, facilitators experienced in guiding scenario development process may prove to be invaluable (see section 5.2.4).

With respect to the number of scenarios to develop, there should not be too many scenarios, as it will be difficult for the audience to keep track of each storyline and see the differences between them (generally more than five scenarios are difficult to handle and comprehend). Developing an even number of scenarios avoids problems with the audience focusing on one scenario that is seen as covering a middle position and thus seemingly being the most likely future. The approach described here leads to four scenarios, which allows some variety while following a rigorous structure. Other exercises may use two scenarios (for example, a best-case and a worst-case scenario) or three scenarios (for example, a reference scenario framed by two alternatives). Still others may aim to highlight the many uncertainties around an issue and use five or more scenarios.

Box 5.8. Building on existing scenarios

Over the last decade a number of scenario exercises were carried out with an environmental focus, such as IPCC (2000), MA (2005a), and UNEP (2002, 2007). When carrying out a new scenario exercise, a first question is whether the new scenarios should in one way or another build on previous ones. And if yes, what should be taken from previous exercises that can either have the same or a very different focus as the new scenarios to be developed.

The answer to such questions depends on a number of factors, such as what the previous scenarios were intended for, what kind of information they produced, and how well they were documented. As a general rule, the more buy in the scenario exercise aims to achieve and the more the exercise is done to help with concrete decisions, the better it is for a scenario exercise to devise its own set of logics and storylines fit to the focal issue at hand.

Nevertheless, existing scenarios can provide a useful source of information and inspiration. The information of previous exercises about specific drivers, for example, may help outline future trajectories in new scenarios. More caution is necessary when simply "borrowing" full scenario logics, however, as a key value in any exercise is to discuss the specific uncertainties to its focal issue and to construct scenarios around these.

In some studies, existing sets of scenarios have been used to frame more detailed thematic analyses. The ATEAM project (Rounsevell et al. 2005) offers an example: The IPCC SRES scenario logics (IPCC 2000) provided a backdrop for regional quantification of land use changes, using appropriate models to further "enrich" these global scenarios at the European scale.

5.3.4 How to describe scenario assumptions and scenario storylines (Stage 3)

The third stage involves developing a set of stories about the future about how various drivers might interact and unfold in different ways based on the scenario logics. Each scenario's logic inherently entails a specific set of assumptions about the drivers and how they develop over the chosen time horizon. In order words, the scenario logic determines the way the events in each scenario play out.

In order to develop illustrative scenario storylines, developers should draw on whatever tools are available to stimulate creative thinking that generates interesting and even provocative descriptions of the future while ensuring that these remain plausible and consistent with the understanding of socioeconomic and environmental processes. Storylines are the basic qualitative descriptions of each scenario. They may, for example, be developed as a series of fictitious events that illustrate how the world might develop over the time horizon and how these events influence the decisions of different stakeholders. This can be told in the form of stylized facts about the future, "letters from the future," short stories that play out in the assumed future world, fictitious future newspaper articles, essays about future prospects, or illustrations of future developments—there is no limit to participants' creativity.

However, it is important to ensure that the scenario storylines relate back to the original focal issue identified in Stage 1. Also, the value of each storyline increases if it remains somewhat comparable to the other storylines developed, and features at least some similar elements. This allows a better understanding of the contrasting pathways the scenarios depict and how individual assumptions shape future trends.

Furthermore, it may be useful to relate the storylines back to how main actors may act under the respective scenarios. If, for example, an analysis of the key factors and actors has been carried out earlier in the assessment, scenario developers can use this information to add additional elements to the storylines and explore different ways for these actors to behave.

Quantifying the main trends in the scenario storylines can also enhance the descriptions of future pathways and may help ensure internal consistency of the respective scenarios. A useful tool to enrich scenario storylines with quantitative information is to run simulation models to quantify future trends of drivers—if time and resources permit and if the issues at hand lend themselves to modeling approaches. Thus assumptions on key driving forces (such as population, economic growth, consumption patterns, lifestyle choices) may be quantified, and possible outcomes for ecosystems and their services (such as food production, climate change, or water availability) can be calculated. In such cases, each model run needs to be based on specific assumptions about drivers' trajectories and their interactions that correspond to the respective scenario storylines. In complex models, however, simulations can take a few months to complete (see also section 5.4).

The translation of qualitative descriptions into quantitative assumptions, however, is not trivial. It is a particular challenge to ensure that this happens in a transparent manner—and all too often the qualitative description and the quantitative analysis are separate processes. An approach used in several previous assessments was to develop tables containing the main characteristics of the storylines and graphic storyline summaries (for example, with upward or downward arrows to indicate important trends; see MA 2005a). This formed a basis for arriving at model input assumptions for the quantitative part of the scenario exercise. Elsewhere, key input parameters were "semi-quantified" using simple ranking or scaling methods (see Box 5.9). And in some cases it may be useful to build on existing scenarios and their quantifications, as mentioned earlier. As noted, this translation of the qualitative descriptions into quantitative information (that can then provide input to further quantitative analysis, see section 5.4) can be done by individual modeling groups, but it is probably best done in close interaction between modeling groups and storyline developers.

Several iterations between qualitative and quantitative techniques help achieve consistent scenarios. But just using either only qualitative methods or only modeling approaches can also generate interesting plausible futures that can stir up discussion. The approach best chosen depends very much on the purpose of the scenario's exercise and the available resources (see also reflections on the balance between qualitative and quantitative information in sections 5.2.2 and 5.4.1). Developers also need to consider the multiple geographical scales that the scenarios can or should address (see Box 5.10).

The final stage in a scenario exercise involves analyzing the implications of the scenarios that have been developed for informing decisions taken today or in the near future (i.e., Stage 4). For this the scenarios have to be "locked in" at one point in time—that is, the stories will not change any more after the iterative process. This is needed to ensure some degree of consistency and comparability in the analysis. Indeed, much of the value of the scenario exercise lies in being able to compare different outcomes, as described in more detail in section 5.4.

Box 5.9. Translating storylines into model input (some examples)

The Millennium Ecosystem Assessment (MA 2005a) provides an example of how scenario storylines can be translated into model input. The four global-level scenarios developed in the MA differ in economic growth and international collaboration (see table). One stylized fact, for example, postulated that high income levels are associated with low fertility levels and low mortality levels. Based on the description of income levels and international distribution, estimates of fertility and mortality in different parts of the world were made. These then were used as model input, so that population projections could be derived (see figure).

Assumptions underlying MA population trends

Variable	Scenario 1: *Global Orchestration*	Scenario 2: *Order From Strength*	Scenario 3: *Adapting Mosaic*	Scenario 4: *TechnoGarden*
Fertility	D: low I: medium	D: high I: low	D: high I: low until 2010, deviate to medium by 2050	D: medium I: medium
Mortality	D: low I: low	D: high I: high	D: high I: high until 2010, deviate to medium by 2050	D: medium I: medium
Migration	high	low	low	medium

D = developing countries; I = industrial countries

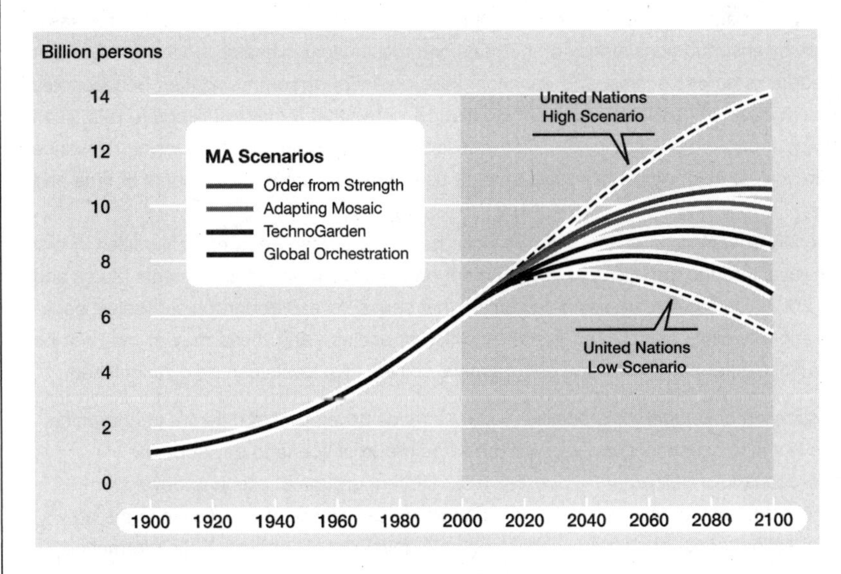

Box 5.9 Figure. Population trends in the MA.
Source: MA 2005a

(continued)

Box 5.9. (continued)

In the PRELUDE project (EEA 2007), a scenario exercise was carried out to stimulate strategic discussions of Europe's land use policies. Considerable attention was paid to transparency in the development of five scenarios that represent different perspectives of Europe's future. Scenarios were constructed around both qualitative and quantitative information and were refined and revised several times. As an important step, key driving forces were quantified roughly using a simple scale from 1 to 10 during the scenario development process, in order to give modeling teams an indication of the relative trends in each scenario.

It is important to note that the opinion of a limited group of scenario developers may play a key role in this interpretation process. One way to reduce the risk of an unwanted determination of outcomes by scenario developers has been the explicit use of worldviews. This approach has, for instance, been successfully applied in the TARGETS project to make scenario developers aware of consistent interpretations of facts different from their own worldview. As a result, rather broad assumptions were made for scenario inputs—and in some cases, models were even adjusted to better reflect the worldviews (Rotmans and de Vries 1997).

Box 5.10. Linking scenarios across multiple geographical scales

Scenario exercises need to consider how to treat the geographical scale(s) of the issues that are investigated as part of the assessment. Increasingly, these exercises address the question of how to build a multiscale scenario (i.e., various geographical scales and cross-scale interactions are explicitly addressed in a single scenario exercise) or multiple scale scenarios (i.e., scenarios are built at different scales independently and then linked with each other).

Linking scenarios across scales has a number of advantages for an ecosystem assessment, such as ensuring consistency of methods and results across scales, enabling the analysis of feedbacks across scales, and enhancing the potential of communication between key stakeholders at different scales. But it should also be noted that it may entail some risks, such as achieving arbitrary consistency (or "over-consistency") of results and thus losing relevance at specific scales. Also, multiscale assessments usually require large investments of time and resources.

Generally speaking, scenario exercises can be linked across geographical scales in two principal ways: through the scenario development process or the scenario elements (Zurek and Henrichs 2007). During an assessment exercise that spans several geographical scales, separate scenario exercises can be carried out at different scales—and these may or may not be coupled across different regions and geographical scales. Five levels can be distinguished:

- *Joint scenario development processes* (i.e., scenarios developed at different geographical scales in a joint scenario exercise, with the same group of scenario developers).
- *Parallel scenario development processes* (i.e., different groups of scenario developers building scenarios at different scales but in more or less parallel processes in terms of focal question, conceptual frameworks, scenario development approach, or information sources).

Box 5.10. (continued)

- *Iterative scenario development processes* (i.e., developing draft scenarios at one scale that provide a starting point for scenario development at another scale, which then provides input and feedback for revision of scenarios at the original scale).
- *Consecutive scenario development processes* (i.e., a set of scenarios first developed and finalized at one geographical level and then scaled to another geographical level).
- *Independent scenario development processes* (i.e., separate exercise carried out at two or more geographical scales—which may or may not inform each other in an informal manner).

Alternatively, scenarios developed at different geographical scales can be linked by using the same scenario elements. The degree of similarity of scenarios across different scales can vary and largely depends on how they are developed. Zurek and Henrichs (2007) describe five different ways of linking scenario elements across scales:

- The closest link between scenarios across scales is achieved when scenarios are *equivalent* or congruent across scales and fully share their scenario logics, key assumptions, and outcome.
- If scenarios are *consistent* across scale, they share main scenario assumptions, driving forces, and trends, but these may play out differently with regard to the scenario implications and outcomes.
- *Coherent* scenarios follow the same scenario logics across scales—in other words, the scenarios "match." This does not preclude substantial differences with regard to how the scenarios play out in the selection of important driving forces, their major trends, and/or scenario outcomes.
- *Comparable* scenarios may be constructed to be largely independent at different scales, connected mainly by the issue they address and possibly addressing the same focal issue.
- Scenarios may be independent and thus *complementary* across scales—yet this does not preclude selected information from scenarios at one scale feeding into scenarios at another.

The degree to which scenario elements will be linked in different scenarios will also depend to a large extent on the type of scenario process applied (see Table). Thus, at the outset of a multiscale exercise the developers should decide what degree of linkage between scenarios is desirable (i.e., how closely the scenario elements should be linked) and design the process accordingly. An example of linking scenarios across scale is the work done by the subglobal groups in the MA. Details on how the process worked for each group can be found in the MA subglobal volume (MA 2005b).

(continued)

Box 5.10. (continued)

Relationship between scenario development processes and scenario outcomes across geographical scales. While any process can lead to any degree of linkage, some are more likely than others.

	Joint process	Parallel process	Iterative process	Consecutive process	Independent process
Equivalent across scales	**very likely**, if a denominator for S1 & S2 exists	**unlikely**, requires rigorous S1 & S2 coordination	**possible**, with unifying S1 & S2 sessions or models	**likely**, if S1 defines reference input / data for S2	**very unlikely**, no coordination between S1 & S2
Consistent across scales	**very likely**, as S1 & S2 developed by same group	**possible**, but only if both S1 & S2 explicit aim for it	**very likely**, if S1 or S2 incorporate respective inputs	**very likely**, if S1 sets binding boundaries for S2	**unlikely**, only if S1 & S2 are de facto consecutive
Coherent across scales	**possible**, if S1 & S2 emphasize different issues	**likely**, if S1 & S2 share starting point, and deviate	**likely**, if S1 & S2 follow same paradigm only	**very likely**, if S1 provides starting point for S2	**unlikely**, only if S1 & S2 are *de facto* consecutive
Comparable across scales	**possible**, if S1 & S2 are developed in parallel de facto	**very likely**, S1 & S2 adopt the same conceptual frame	**possible**, if S1 & S2 aim to address different needs	**likely**, deviation if different focus S1 & S2, same frame	**possible**, if S1 and S2 use similar conceptual frame
Complementary across scales	**unlikely**, as S1 & S2 conceptually independent here	**likely**, if S1 & S2 are parallel yet autonomous	**unlikely**, this implies that iteration fails	**possible**, if S1 provides info only for S2	**very likely**, S1 & S2 address similar issues differently

Note: S1 and S2 denote two separate scenario exercises at two different geographical scales.

Source: Zurek and Henrichs, 2007.

5.4 How to analyze scenarios

Section's take-home messages

- To analyze scenarios, either qualitative or quantitative approaches or a combination of the two can be used. Each has strengths and weaknesses, and using a combination can harvest the advantages of both, but it requires substantial amounts of time and resources.
- In the context of ecosystem assessments, a broad range of modeling tools exists at various scales that can support quantitative analyses—but it is difficult to find a single model that can adequately capture all the dynamics depicted in a scenario.

- An analysis across a set of scenarios—for example, with respect to similar or differing trends—can facilitate discussions about possible future trends and their plausibility, the impact of different response options, or the robustness of different response options under different assumptions.

5.4.1 Analyzing scenarios—combining qualitative and quantitative information

The analysis of the potential implications of scenario assumptions is often core to environmental scenario exercises: much of the value of the exercise lies in being able to compare different outcomes. Such comparisons can reveal unanticipated results and provide different stakeholder groups with insights about the outcomes of the future pathways they may have advocated. Scenarios can be analyzed using a range of different tools—both qualitative and quantitative. Section 5.3 highlighted the role of and interplay between qualitative and quantitative information when developing scenarios. This section explores this notion in further detail, shows how combining these different types of information can be used in the analysis of scenario implications, and discusses the role of analytical tools in scenario analyses in more detail.

As noted, the differentiation between qualitative and quantitative scenarios has proved to be useful. But this distinction is not always clear-cut. Many reference (or business-as-usual) scenarios, for example, focus on projections of expected future changes without emphasizing explicitly the underlying assumptions (see, for example, IEA 2007, FAO 2002, EEA 2005). Conversely, qualitative scenarios about possible future developments are seldom void of all quantification. Thus, as most scenarios implicitly combine qualitative and quantitative information to some extent, this differentiation is mainly useful for highlighting the principal focus of a given set of scenarios.

The specific roles of qualitative information within scenarios include the following:

- To capture the multiplicity of perspectives of actors, and the significance that those actors assign to events. This allows subjective insights from stakeholders to be taken into consideration. The necessity of subjectivity (in understanding social issues) is due to the recognition that there might be several different alternative perspectives of reality, all of which may be "valid" and should be explored. Models are not capable of doing this.
- To enlarge the range of outcomes of a scenario analysis. Models, for example, cannot address all issues at all geographical scales at the same time. In particular, local developments are usually very dependent on close relations between actors that cannot be quantified by models. Qualitative approaches can provide insights into these aspects and therefore can broaden the scope of the scenario analysis by picking up issues that models cannot capture.
- To deal with shocks and disruptive scenarios. Usually simulation models are constructed to provide insight into gradual changes in systems, since the numerical approach can play out its strengths in capturing this type of change. In reality, however, societal changes are usually most dominant when extreme events occur. Where models lack numerical insights in capturing these extreme events, qualitative approaches can dwell upon their potential impact.

The specific roles of quantitative information within scenarios—for example, based on modeling output—include the following:

- To "enrich" qualitative scenarios by showing trends and dynamics not anticipated in the storylines. Models may provide another way to explore the consequences of changing one set of parameters for other parts of the system. In this way, models can provide insight into complex relations of a system that are difficult to grasp. Clearly, models are restricted to aspects that can adequately be translated into numerical concepts.
- To check the consistency and plausibility within scenarios. Models include various constraints on how model parameters may develop. In a climate model, for instance, the rate at which greenhouse gas concentrations can be reduced is constrained by knowledge on the removal rates of these gases from the atmosphere. In that context, models can be used to test the consistency of initial storyline descriptions. It should be noted that if models provide different insights than the original storyline, the model is not necessarily right (as models are themselves only one possible simplification of reality). Confronting "qualitative expectations" and model outcomes might lead to a very useful mutual learning exercise.
- To provide relevant numerical information. If uncertainty about the future allows quantitative assessments, in many cases numerical information can be more convincing than purely qualitative descriptions. For example, for decision makers in the field of climate policy it is useful to get some idea of the costs involved in various climate policy proposals; for decision makers in the area of bioenergy it is important to know how much land is likely to be used in the context of various proposals.

Recently, scenarios that explicitly aim to combine qualitative and quantification information—for example, the "story-and-simulation" approach (see also EEA 2001)—have gained prominence and have been seen as particularly useful in international environmental assessments (see Box 5.11; see also IPCC 2000, Cosgrove and Rijsberman 2000, Alcamo et al. 2000, UNEP 2002, MA 2005a, UNEP 2007). Such approaches make best use of the respective strengths of qualitative and quantitative information and should allow for interaction. For example, the assumptions captured in qualitative storylines may be adjusted on the basis of model outcomes. At the same time, model inputs (or even the models themselves) can be adjusted to the respective scenario storylines. Preferably several interactions are organized between the storyline development and the model-based analysis during a scenario exercise in order to increase the consistency of the two types of information and ensure cross-fertilization of both processes (see section 5.3.4).

The degree of quantification is an important choice (see section 5.2.2). While using models provides some clear advantages, it comes at a cost (in terms of time and money, but also in terms of reduced flexibility in which issues can be explored and how). And for certain issues, models might not be available or appropriate. The degree of quantification also depends on the characteristics of the scenarios that are developed. In scenarios that deliberately explore the implications of singular events (such as shocks or surprises) rather than gradual trends, modeling approaches are harder to implement, as most models have been calibrated against overall long-term trends and struggle to deal with discontinuities.

Box 5.11. Some examples of the role of modeling in developing scenarios

In the Millennium Ecosystem Assessment, qualitative and quantitative scenario elements played an equal role (MA 2005a), providing some examples of the specific role of modeling.

- Some scenarios were designed to include climate policies while others did not. Climate policies will lead to significant changes in the energy system (e.g., less use of fossil fuels with carbon capture and storage and more use of renewable energy). These changes will have numerous indirect impacts. In the MA, energy models were used to illustrate the trends in air pollutant emissions in relation to climate policy—something that cannot be easily done without model analysis.
- Land use trends are determined by factors such as population change, dietary changes, and crop yields. These trends may partly occur independently, but they also interact: increasing land prices as a result of scarcity is major driver behind yield increase. In the MA, it was found by quantitative modeling that offsetting trends in driving forces led to a much smaller range in land use outcomes that originally expected. These offsetting trends include the facts that high economic growth scenarios (with a related strong trend toward meat-intensive diet patterns) coincided with low population trend, and endogenous drivers for yield (land prices) acted as a mitigating factor. Subsequent discussions led to revisions in both the original storylines (weaker differences in land use) and modeling.
- For the MA scenarios, hypotheses were formulated on the consequences of the different scenarios for biodiversity. For instance, it was expected that biodiversity loss in a *Global Orchestration* world may be larger than in a *Technogarden* world. Still, without further quantification these statements are not more than hypotheses—which may be ignored by certain target audiences. Therefore a simple biodiversity model was developed to integrate trends in factors such as land use, climate, and nitrogen deposition and to illustrate the expected consequences for biodiversity of the different scenarios.

There are a few substantial risks, however, in using a combined "story-and-simulation" approach:

- Scenario development with models requires substantial resources and time; as a result, modeling results tend to come in late—reducing the opportunity of interaction.
- One approach may dominate over the other; this obviously reduces the purpose of doing both activities.
- The two outcomes may be inconsistent; Parson et al. (2007), for example, criticizes the Millennium Ecosystem Assessment (MA 2005a) for inconsistencies between storyline and quantification. (It should be noted, however, that it was a deliberate choice to present these inconsistencies in the MA: inconsistencies were first confronted and discussed and those that could not be resolved were presented equally, so as not to introduce bias between qualitative and quantitative analysis.)

Thus, while there are great benefits in combining qualitative and quantitative information within a scenario exercise, this might not always be feasible or desirable —and it depends on the scenario exercise's context and goals. In such cases the strengths of the different types of information should be weighed carefully against each other.

5.4.2 How to analyze the implications of scenarios regarding ecosystems and human well-being

There is no single recipe for analyzing the implications of scenarios. In order to illustrate the principal steps of such an analysis, they are elaborated here as a continuation of the deductive scenario development introduced above—assuming that the main assumptions underpinning the scenarios have been quantified (as indicated in section 5.3.4) and that a "story-and-simulation" type process to scenario analysis is followed.

The implications of assumptions made about future developments under each scenario can be analyzed using both qualitative approaches and simulation models. The term model or modeling tool is used here to refer to tools that describe in a formal manner the relationship between different elements that together determine the behavior of a larger system. In other words, models are geared toward linking assumptions on driving forces to the possible implications these may have in a quantitative manner. Note that this section focuses on the use of such modeling tools to support scenario analyses—while recognizing that a host of nonquantitative tools also exist to analyze scenario implications in a more qualitative manner.

At the core of any ecosystem-related scenario analysis is the exploration of the future trends of key driving forces, their interactions, and their implications for ecosystems and human well-being. In this context, it is useful to distinguish between different types of driving forces—direct drivers (such as land use change, climate change, or air pollution) and indirect drivers (such as demographic, economic, or technology development, as well as human behavior or institutional factors) (see Chapter 4). For many of these drivers of ecosystem change a variety of detailed models exist at most geographical scales. Such models can be incorporated into an assessment. Examples of issues for which simulation tools that have been used in past assessments at a global scale presented by the Millennium Ecosystem Assessment (MA 2005a) or the Global Environment Outlook (UNEP 2002, UNEP 2007) include demographic trends, economic development and associated changes in demand for physical goods, emissions, food demand and production, climate change, land use change and land cover, and water resources. For many other issues, however, modeling tools seem to be scarcer (and arguably cannot exist, due to the high degree of indeterminism in socioeconomic developments). This, for example, refers to sociopolitical drivers, educational drivers, and cultural drivers as well as trends in science and technology.

With regard to the interaction between drivers, currently available quantitative approaches can be somewhat limited—and qualitative methods are often needed to complete a scenario analysis. For many of these linkages, "stylized facts" can be drafted that might be commonly accepted but are not part of formal models. In some cases, this is because models simply do not capture these links, as most models only capture one specific driver. In other cases, the stylized facts are still too weak to use in a quantitative way in models. Examples of such stylized facts include assumptions used in the MA scenarios that "fertility rates are low under high economic growth scenarios" (based on the fact that current fertility levels are low in high-income countries) and that "technology development is high under globalization scenarios." Both these stylized facts have been included in the scenarios by introducing them into the scenario storylines. For the models, they simply constitute exogenous assumptions.

In the case of assessment of complex systems such as those related to ecosystem services, it is unlikely that one single model can be used to describe the full system. Although some highly aggregated models have been developed that specifically try to focus on the overall picture (for example, GUMBO [Bouwmans et al. 2002] or IFs [Hughes 1999]), such models often lack the detail that is needed for a policy-relevant assessment. Therefore, in recent ecosystem assessments a suite of models has been used. Usually, the different models are harmonized by using the same driving forces and by linking the output of one model to the input of another model. In some cases the modeling linkage is even more sophisticated, by allowing different models to iterate the model outcomes. In this way, the benefits from the specific (expert) knowledge included in each model can be harvested.

At the same time, however, the potential for inconsistencies grows when scenario exercises use multiple models and attempt to harmonize them, particularly when some key quantities are externally specified for some models and calculated within others. These inconsistencies can be minimized, however, for example by iteration across models (see, for example, Box 5.12) or by choosing model links at the point of limited interaction (energy models and emissions, for example, are relatively independent of changes in climate). Nevertheless, it should be noted that attempts to avoid such inconsistencies by standardizing model outputs can entail more serious risks by obscuring interpretation of results and precluding use of model variation to illuminate uncertainty. Also, attempts to connect qualitative and quantitative aspects of scenarios have been particularly challenging for pursuit of consistency. Different narrative scenarios often reflect different assumptions about how the world works, which correspond more closely to different model structures than does parameter variation. Better integrating the two approaches will require developing ways to connect narrative scenarios to model structures rather than merely targeting values for a few variables that models are then asked to reproduce.

Analyzing response options also plays a key role in most scenario assessments. Sometimes these options are to some degree incorporated into the scenario assumptions—but in most cases it is far more useful to analyze possible response options in a separate step against the backdrop of different scenarios (and thus test the robustness of various options). Response options can be analyzed in a range of different manners (see also Chapter 6).

One option to analyze the effectiveness of different response options is to use a model to analyze various options against a single criterion (such as cost minimization or meeting specific targets). An example of this is the standard practice in climate mitigation analysis: by attaching a price to greenhouse gas emissions, measures that reduce emissions are induced in the model. Analysts report the measures that are chosen and the associated costs. The analysis can be done starting from different nonpolicy scenarios in order to map out the consequences of uncertainties in driving forces (see, for instance, van Vuuren et al. 2007).

An alternative approach is to use scenario storylines to assess whether a certain measure fits into the set of assumptions that depicts future world settings. For example, in the EU-Ruralis project (Westhoek et al. 2006) different agricultural policies

Box 5.12. Coupling models: experiences from the MA and EU-Ruralis

In the Millennium Ecosystem Assessment (MA 2005a), quite a large set of models were coupled to describe the broad range of changes in drivers of ecosystem services. They were "soft-linked" in the sense that output files from one model were used as inputs to others (see figure). The time interval of data exchanged between models was usually one year. The model linkages included factors such as food production, climate, irrigated areas and water consumption, electricity use and livestock, river discharge, land use change, and nitrogen deposition. Taking the example of food production, information was forwarded from the agro-economic model IMPACT to the integrated assessment model IMAGE to compute land use and land cover changes at grid level that were consistent with the agricultural production computed in IMPACT.

In the EU-Ruralis project (see Westhoek et al. 2006) a modeling framework was constructed to improve the linkage between an agro-economic model (GTAP) and IMAGE. Iterations were done between GTAP and IMAGE to increase the consistency of the information on agricultural production, availability of land, and climate change. After the iteration, a consistent set of modeling results was obtained from GTAP and IMAGE, allowing analyses on agricultural consumption, production and trade (from GTAP), and land use change, greenhouse gas emissions, and climate change (from IMAGE).

Linkages between models are seldom straightforward and usually require the upscaling (aggregation) or downscaling of various data. The IMPACT model, for instance, uses a more detailed regional disaggregation level than the IMAGE model. In almost all cases, however, regional scaling was possible by combining regions (upscaling) or by assuming proportional changes (downscaling). The models were not recalibrated on the basis of the new input parameters provided by the other models, but in most cases the models had been calibrated using comparable international databases. The new linkages therefore did not lead to major inconsistencies in assumptions between the models.

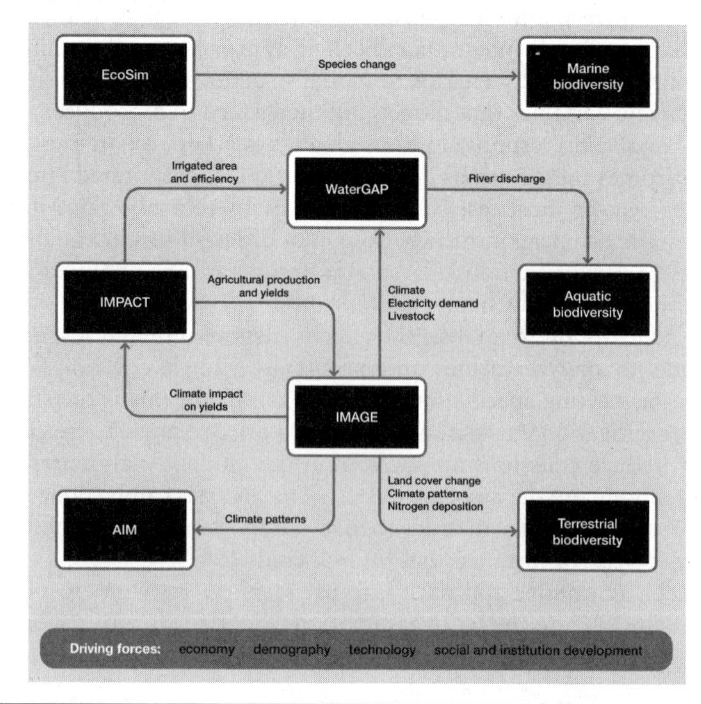

Box 5.12 Figure. Linkages between models in the MA. *Source:* MA 2005a

are added on top of the climate change scenarios laid out by the IPCC (IPCC 2000) in a manner consistent with the underpinning scenario logics. Agricultural policies in this example are differentiated between full liberalization of the Common Agricultural Policy of Europe in a globalized and economy-focused world (in the A1 scenario), a complete continuation of current policies under a regionalized and economy-focused world (in the A2 scenario), and even an increase in agri-environmental payments in a regionalized and environment-focused world (in the B2-scenario).

Obviously, the starting point (the so-called nonintervention scenario) plays a major role in response analysis. For instance, in analyzing the climate response assumption, the technology assumptions in the nonintervention scenario play a major role. Slow technology development—and thus high emissions—may imply that certain targets are unachievable. As a result, these nonintervention scenario assumptions often remain an important part of discussion (see, for example, Pielke et al. 2008).

5.4.3 How to treat uncertainty in scenario analysis

While scenario exercises themselves form a tool of uncertainty analysis by exploring different possible futures, likelihoods and uncertainties can also play a role within scenarios (addressing, for example, questions such as "how certain is it that a certain assumption made in a scenario will come true?"). A model-supported scenario analysis can usually only grasp a few key uncertainties by varying among scenarios. It should be noted that complex narrative scenarios pose special problems in representing and communicating uncertainty, given the large number of posed interactions across parameters.

At the same time, many users of scenarios would like to be presented with some indication of the possible range of outcomes, or even the likelihood or probabilities associated with future trends. Most assessments make a strong case trying to prevent such interpretation. Constructing scenarios offers a way of thinking about the future in a structured manner, and the aim is to provide a range of outcomes in which the probability is not the most important aspect. In the MA (MA 2005a) and IPCC (IPCC 2000) scenarios, for example, it is emphasized that no probabilities could be attached to any of the constructed scenarios. This has often been interpreted as equal probability—but in fact this interpretation is incorrect.

However, current scenario practice leaves the question unresolved of whether and how to best include probabilistic information (Groves and Lempert 2007). It has been reasoned that assigning probabilities to the likelihood of different scenarios would be misleading and would undermine the credibility—not least due to the overwhelming influence of societal choice (and associated "deep uncertainty") (IPCC 2000). Elsewhere, particularly in the context of climate change impacts, it has been argued that policy analysts and decision makers need probability estimates to better assess risks and decide how to respond to them (Schneider 2001, Schneider 2002, Webster et al. 2002). Several studies have applied the contrasting probabilistic approach to emission scenarios (Webster et al. 2002, Webster et al. 2003, Richels et al. 2004). An important critique formulated against this approach is that attempts to assign subjective probabilities in a situation of ignorance forms a dismissal of uncertainty in favor of spuriously constructed expert opinion (Grübler and Nakicenovic 2001, Grübler et al. 2006).

Instead, scenario analysis often explicitly aims to consider futures that span a range of possible outcomes—making probabilities less relevant. Still, there seems to be a natural inclination to interpret the range of scenarios as an indication of possible outcomes. For instance, the Millennium Ecosystem Assessment outcomes (MA 2005a) for loss of natural area range from a 20% loss (under one set of assumptions) to a 10% gain (under another set). This can easily be interpreted as an indication that the most likely development is a change in area between –20% and +10%. In most cases such interpretation is incorrect, as the existence of one low scenario does not exclude the possibility of even lower ones. However, given the need to explore the range of possible outcomes, it is worthwhile to consider scenario development against the backdrop of extreme cases. And most assessments in fact do so.

It may be worthwhile to try to combine the strength of scenario analyses with probabilistic approaches. In both the IPCC SRES report (IPCC 2000) and the MA (MA 2005a), for example, the assessment first discusses the full range of possible outcomes for different scenario drivers—and, in case of the IPPC, also scenario outcomes. Next, the assumptions made for the different drivers are related to the possible range of outcomes. For example, a scenario that emphasizes high population growth would use an estimate close to the upper range, while a scenario that emphasizes low population growth would use an estimate close to the lower range. As a result, the full scenario range would somehow represent the (uncertainty) range found in the literature. To some degree, a similar situation exists with respect to scenario outcomes. If scenarios would all represent a small range of possible outcomes for some key variables, they would fail to communicate the possible range of outcomes. For these variables, therefore, a fuller scenario range is most helpful.

5.4.4 How to analyze across a set of scenarios

There are a number of ways to compare outcomes of different scenarios (see Table 5.4). In general, lessons can be drawn from focusing on either the similarities or the differences in trends across the scenarios. These can be connected to the policy choices made along the different pathways. Lessons for decision making can also be drawn by comparing the risks taken and the benefits gained by different groups of society or by mapping out the trade-offs in each scenario.

Insights emerge from questioning the assumptions made within one story or pathway and comparing its outcomes with another possible pathway into the future. This analysis can clarify what is known and what is uncertain about the future. It also sheds light on unexpected results of a particular pathway. In other words, the scenario analysis can reveal uncertainties and help decision makers avoid the unintended consequences that often plague policy processes.

Most assessments produce a set of scenarios in order to explore possible developments under different sets of assumptions. Here, the set of scenarios can be used to explore possible future outcomes—and such outcomes may include similar trends and differing trends, as well as offsetting trends.

In the case of differing trends, scenarios can depict possible different future situations and thus communicate the consequences of different assumptions or trends. For example, greenhouse gas emissions under the Millennium Ecosystem Assessment's *Global Orchestration* scenario are very different from those under the *Technogarden* scenario. This emphasizes how assumptions about the future dynamics with regard to economic growth, material consumption, and technology development are of critical importance for future trends in greenhouse gas emissions. In

Table 5.4. Options for Comparing Scenarios

Options for comparing across the scenarios	Example from the Millennium Ecosystem Assessment scenarios
Look for future developments that are the same in all scenarios	Same trend of rising world population up to 2050 in all scenarios, then stabilization; exact population numbers in 2050 differ.
	Global forest area declines up to 2050 in all scenarios: velocity of trends differs.
Look for uncertain future developments that differ across scenarios	Number of malnourished children in 2050 differs widely among scenarios.
	Quality and quantity of available water resources by 2050 differ widely among regions and across scenarios.
Identify trade-offs described in the scenarios	Risk of trading off long-term environmental sustainability for fast improvement in human systems (*Global Orchestration*).
	Risk of trading off solutions to global environmental problems (requiring global cooperation) for improving local environments (focusing on local solutions only) (*Adapting Mosaic*).
	Risk of trading off biodiversity conservation for food security (*Global Orchestration*).
Identify policy options that make sense in all scenarios	Major investments in public goods and poverty reduction, together with elimination of harmful trade barriers and subsidies.
	Widespread use of adaptive ecosystem management and investment in education.
	Significant investments in technologies to use ecosystem services more efficiently, along with widespread inclusion of ecosystem services in markets.

Source: WRI 2008.

their combination, the two scenarios are indicative of both high emission and low emission scenarios published elsewhere.

Similar trends for outcome variables—if the outcomes of very different assumptions about underlying drivers are similar or identical—might be just as indicative. Such similar trends may occur for very different reasons, including similar consequences of different trends, offsetting trends, and delays. In the IPCC SRES scenario set (IPCC 2000), for instance, the issue of similar consequences of different trends is illustrated by the B1 and A1T scenarios. Both scenarios depict similar greenhouse gas emissions—but for very different reasons. In the B1 scenario, low emissions are caused by a preference for renewable energy and environmentally friendly lifestyles (in combination with favorable technology assumptions). In A1T, however, the only driver is the fast development of alternative, carbon dioxide neutral technologies.

The issue of offsetting trends plays some role in the outcomes of the Millennium Ecosystem Assessment scenarios for land use. In most agroeconomic models, a major driver for technology development for yield improvement is land scarcity. As a result, the strongest pressure to improve yields exists for a scenario with (initially) high land scarcity, while the lowest pressure exists for a scenario with (initially) low land scarcity. In other words, drivers that result in diverging land use (such as population or diets) are partly offset by this "equilibrium-seeking" mechanism. This is an important reason that the outcomes regarding land use in these scenarios are much closer together than the original population, income, and lifestyle assumptions would suggest. Similarly, the impact of inertia can be observed in the trends in global mean temperature increase in the same scenarios: in 2050, despite very different greenhouse gas emission trends, the range of temperature outcomes is still relatively small. The main reason for this is the inertia in the climate system, which does not allow global mean temperature to differ much across scenarios in the short term.

Another way of using scenarios in the analysis of response options is the so-called robustness analysis. Here, the focus is not on finding optimal responses based on a single scenario but on identifying response options that will work against a range of different scenarios and, as a result, will be more robust regardless of which scenario is assumed to unfold (see, for example, Lempert et al. 2004).

It should also be noted that the analysis needs to be fit for its purpose with respect to its scale. Scenarios are mostly designed to explore long-term futures using models that focus mostly on long-term dynamics. As a result, long-term scenarios in most cases have only limited variation in the short term. This, however, does not imply that short-term ranges in output variables are small. This range, however, will result from very different dynamics. For instance, while in the long term energy prices are mostly driven by technology development and depletion, in the short term factors like underinvestment, short-term variation in economic growth, and speculation may play an important role.

It is worth noting that not all model results can be used in the same way to present the outcomes of a scenario analysis. While some results may be presented with relatively high confidence and can serve as backdrop to detailed quantitative analyses, other outcomes should be seen as purely indicative and lend themselves primarily for comparison purposes. A good example of an indicative indicator is land use. Clearly, land use maps should not be used for predictive purposes, given the large uncertainties that exist regarding future land use. Therefore, land use maps should be communicated in a way that justice is done to the uncertainties that surround the results. In the EU-Ruralis project (Westhoek et al. 2006, Eickhout et al. 2007), for example, land use maps are used to pinpoint areas where abandonment occurs in all scenarios. In this case it is assumed that when land abandonment occurs in all four scenarios, an area is susceptible to land abandonment irrespective of the scenario assumption.

This process of comparing and analyzing the scenarios then leads into a discussion on response options to meet future challenges. How the scenarios can enhance the (policy) debate on how to tackle the identified problems and along which pathways to proceed depends to a certain extent on who the scenario developers were, on the authorizing environment, and how the scenarios are written up and disseminated. For example, scenarios built for the purpose of decision making or strategy development can be more effective if the key stakeholders taking the decision are included in the scenario development process, so as to trigger a discussion and a reassessment of views and perspectives on the decision to be taken. To stimulate a wider debate on issues at stake, the dissemination of the scenarios can also be carried out not just via reports or books, but also through such other means as theater plays or cartoons about the scenarios, movies about them, or other creative, artistic ways to illustrate the stories, as described in the next section.

5.5 How to use and communicate scenarios

Section's take-home messages

- Scenario exercises do not end after developing and analyzing a set of scenarios; their use and active communication is as important as the scenario development process and needs to be planned well ahead.
- Effective use and communication of scenarios depends on their respective contexts and, in particular, on whether the scenario process applied was appropriate, the type and format of scenarios were suitable, an adequate communication strategy exists, and the scale of analysis fits the context.
- Key features of effective scenarios depend on their main purpose: in order to support scientific exploration and research, credibility and transparency are essential; for education and public information, scenarios need to be relevant and accessible; to be effective as decision support and in strategic planning, scenario exercises need to address relevant current concerns in a legitimate and challenging manner, which is why the focus is on the process as much as on the scenarios themselves.

5.5.1 The main dimensions of scenarios use

Developing and analyzing well-crafted scenarios is a first step toward getting a broader, informed overview on present and future developments, available options for action and their potential effectiveness and robustness. Translating findings from a scenario exercise into effective action is the next step. However, experience shows that this is not a simple or straightforward process (Wilson 2000). The context dependency of using scenarios makes it difficult to establish a standard recipe for success.

Indeed when the aim is to support decision making or trigger learning processes, it is difficult to determine and measure success. Is it enough, for example, if scenarios trigger lively discussions for a day? Or should they lead to longer-term changes in behavior, which are hard to measure? Can it already be seen as a success if policy makers start to think about longer-term consequences and if scenarios help to better manage conflicts between policy makers and key societal stakeholders? Or has a scenario exercise only been successful if it really has an impact on decision-making processes? If so, what is the appropriate time perspective for a successful uptake of a set of scenarios in the context of decision making?

The ambiguity of answers to these and similar questions underlines the importance of defining clearly the purpose and goals of a scenario exercise at its outset—as stressed in section 5.2—and to scrutinize the important factors determining successful use. A number of influential factors in this regard should be carefully checked.

A first factor is the appropriateness of scenario development process. Whether scenarios can be used successfully is to some extent predetermined by their development process. If, for example, scenarios are to be used in support of scientific exploration or research but are not perceived to be sufficiently credible and fail to conform to criteria of good scientific practice, achieving success will be difficult. The perception of saliency, credibility, and legitimacy (as discussed in Chapter 1 and section 5.1) is thus an important influence on the success of scenario exercises. (Sections 5.5.2 to 5.5.4 discuss in greater detail which of these characteristics are of particular importance in each context of scenario exercises and which components thus need particular attention.)

A second factor is the suitability of the type and format of scenarios to their context and intended use. A normative scenario exercise, for example, is not well suited when trying to understand the implications of certain decisions depending on possible future evolutions. Conversely, this type of scenario is well suited to helping start a broader discussion about the prospects of initiating long-term policy developments to reach a certain target or the work toward changes to current structures desired by societal groups.

A third factor is the match between context and communication strategy. Here an analysis of the context within which a set of scenarios will be used is essential: What is the key target audience, and what kind of communication strategy and presentation style is suitable for it? A public audience, for example, needs a different presentation format than a scientific audience: triggering lively discussions with a diverse set of societal stakeholders requires a compelling, attractive presentation of alternative storylines and their consequences, whereas it is not too important to present methodological details. The opposite might be true for a scientific audience. Furthermore, the time requirements might differ: Initializing a strategic conversation among different stakeholders needs at least a day to present all the scenarios in a circuit and give participants time to get into the modus of thinking in alternatives.

Identifying key target audiences and defining the suitable communication strategy requires a good analysis of the overall context and corresponding institutional landscapes: What are the outstanding questions, or problematic issues, the target audience is facing? Is the topic already well established on the agenda, or is it emerging? Is the target audience long-standing actor networks with well-established belief systems or loose, open networks? What are the time resources and the level of knowledge? Any communication strategy should include clear criteria that visualize successful achievements. This will help to monitor and evaluate process. The criteria should allow for flexibility to adapt in the process but should set minimum standards (such as, for example, engaging relevant policy makers in discussions about the full set of scenarios in a workshop or meeting). Visualizing a successful outcome helps people understand the aim of the exercise.

A fourth factor is the fit of (geographical) scale. Whether scenarios can be put to effective use is also often a question of addressing the appropriate scale and whether the scenarios address and display trends at the right level—that is, the level of interest and relevance to the intended audience or political decision makers. For example, national government officials are first and foremost interested in learning about the implications for their own country under the respective scenarios.

The remainder of this section discusses in greater detail the conditions of using scenarios for scientific exploration and research, for education and information, and for decision support and strategic planning.

5.5.2 How to use scenarios for scientific exploration and research

Within scientific assessments, scenarios often provide a structure for the analysis of alternative future developments. In addition, scenario-based approaches that combine storyline development with modeling may give indications for those inputs that cannot be explicitly calculated within the respective models but that are nevertheless necessary for representing full accounts of plausible future developments (i.e., they specify exogenous inputs to a model).

The products of a scenario exercise—the actual scenarios—can also be used as an input into other assessments (Parson et al. 2007). Indeed, to date this appears to have been the dominant use of scenarios in activities related to international ecosystem assessments (see, for example, the global MA scenarios [MA 2005a] or the IPCC SRES scenarios [IPCC 2000], both of which have been used in numerous projects as a source of input by subsequent assessments such as Reidsma et al. [2006] or Rounsevell et al. [2005]).

In addition, scenarios can also be used as input into other research activities or as a platform to organize inter- and intradisciplinary discussion and exchange. Scientists external to the scenario development process might be asked to comment on the scenarios, to highlight gaps in the analysis, and to indicate needs for further improvement. In the end, this can contribute to establishing an agenda of long-term research needs with regard to ecosystem assessment.

Appropriateness of scenario development process

If scenarios are to be successfully used for scientific exploration and research, they first and foremost need to be perceived as credible according to standards of good scientific practice. Being considered as legitimate and salient is, of course, helpful but not as important. A note of caution is warranted, however, with regard to the credibility of scenario exercises. To some extent *all* scenarios are speculative and will never fully comply with standards of good scientific practice (in particular, there is considerable discussion about whether scenarios are reproducible or can be validated). Thus scenarios will never be based on scientific knowledge alone but always need to blend different forms of expertise and judgment. Therefore transparency regarding the process, the underlying reasoning assumptions, and the methodology used—but also plausibility with regard to defining assumptions, choosing data, and applying models—are of critical importance in order to ensure credibility (see Parson et al. 2007 for this discussion).

Suitability of scenario type and scenario format

In the context of scientific assessments, scenario exercises need to comply with the overall format requirements of the wider assessment process it is supposed to feed into (see also Chapters 1 and 2 as well as Box 5.2). A qualitative scenario exercise will hardly provide relevant information into an otherwise purely quantitative assessment. Past assessments have found the combination of qualitative and quantitative information to be useful. Furthermore, the assumptions and methods used need to be transparently documented and to be congruent with the overall basis assumptions of the assessment exercise.

Match between type of context and communication strategy

As scenarios provide a structure for future analysis or can be used as input into other assessments, the key target groups in this context are often other scientists. Analyzing the political and institutional context to find the relevant target groups is therefore not too important. But it is important to understand the factors that shape scientific discourses in the relevant communities, even—or especially—if the aim is to enrich the discourse or challenge dominant paradigms.

For a scientific audience, the communication should be geared toward the scenario outcomes, as well as the data and methods that have been used in order to allow a transparent discussion of key assumptions and model applications. Publishing in peer-reviewed academic journals is often a prerequisite for information to be acknowledged as scientifically sound and trustworthy. The related requirements—especially with regard to transparency about the methods applied and documentation of underlying assumptions—need to be taken into account when designing scenario exercises that often cut across disciplinary boundaries.

More and more frequently, the information gained from scenario exercises is compiled in multimedia presentation tools (be it Web-based or a CD-ROM) that allow accessible documentation of the wealth of information and methodological assumptions as well as much of the underlying data. Examples here are the presentation tools for two recent scenario exercises focusing on land use in Europe—that is, EU-Ruralis (see www.eururalis.eu) and PRELUDE (see www.eea.europa.eu/prelude). In most cases, such tools should not focus on the exact numbers but on the comparison of results between scenarios or regions. One way to do this is via "classical" graphs or tables. Different presentation methods have been used and developed in scenario analysis, including vague graphs, traffic lights, and spider diagrams. (Some examples are provided in Figure 5.6). In a spider diagram, for example, different scenarios or regions can be displayed in one figure, offering an analysis of the difference between those scenarios or regions. Usually the focus of such analysis is not on the exact numbers but on the relative differences.

Fit of scale

As described in section 5.3, various methods help link scenarios across different scales. While a congruence of scale is of course helpful for successful use of the scenario's results in an overarching assessment or another study, it is not a precondition. It is more important to have a transparent outline of the key assumptions used, input data, and methodological approaches at the respective geographical scale to determine the relevance of the findings for other research purposes. If the scenarios are to be used to inform long-term research programming, however, the issue of "fit of scale" becomes more relevant.

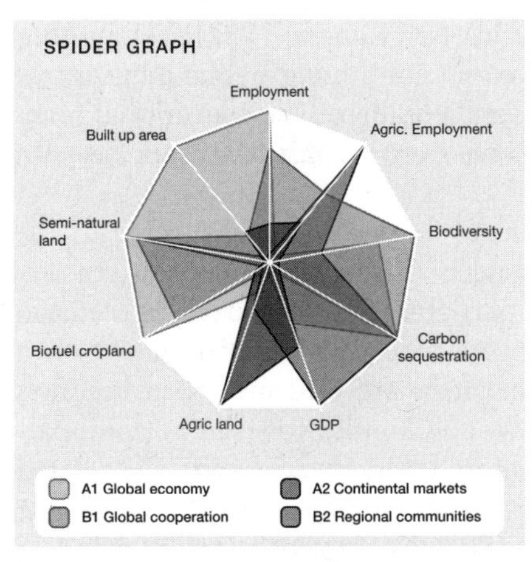

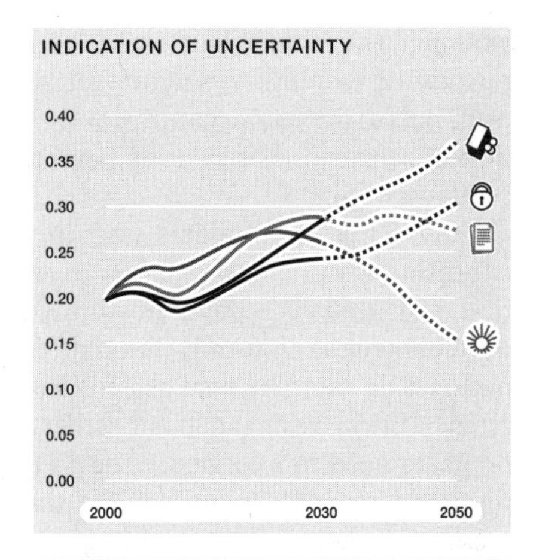

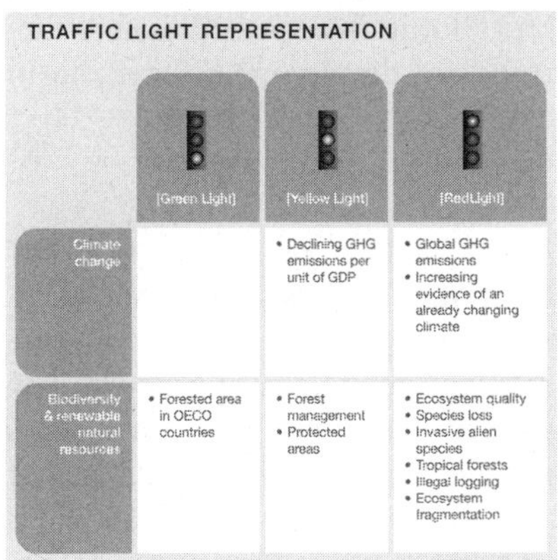

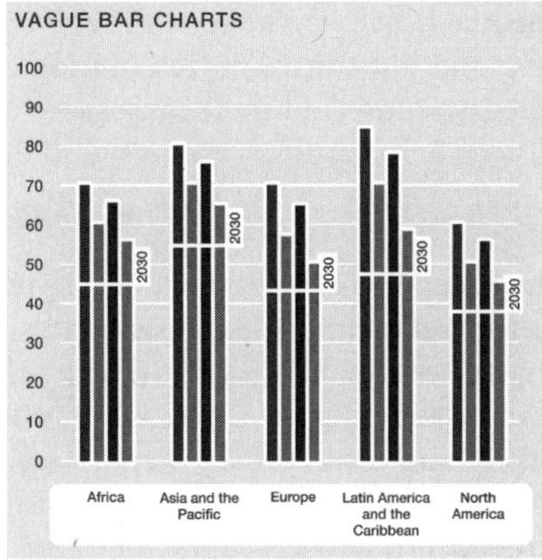

Figure 5.6. Different presentations of scenario output from EU-Ruralis study, OECD Environmental Outlook, and GEO3.
Sources: Rienks 2008; OECD 2008; UNEP 2002

5.5.3 How to use scenarios for education and information

Scenarios are often used to provide policy makers, stakeholders, and the general public with a structured overview of plausible future developments—that is, they offer an approach to making sense of the myriad of future-related information, helping to create a common language and a common platform for diverse communities to discuss their respective perceptions of the future. Thus scenarios can contribute to the framing of the key choices, risks, and opportunities that organizations or societies face (Starr and Randall 2007). In other words, they have an informational function.

Quite often, this is combined with an educational function. Policy makers, stakeholders, and the general public often ignore or deny the possibility of far-reaching change and stick to successful decision modes of the past. Scenarios can help break such positions of denial by illustrating the plausibility of different trend developments or the consequences of certain decisions or inaction. Normative scenarios can also be used—and have been used—to frame wider public discussions about desirable futures that contribute to processes of fundamental social change, as, for example, the Mont-Fleur scenarios in South Africa (see Kahane 1992). Yet another purpose is to build capacities for scenario development among public administration, public interest organizations, or corporate stakeholders. The educational function of scenarios has steadily gained in importance over the last few years (see van Notten 2005).

Involving stakeholders into the process of developing scenarios increases the educational value and collaborative learning aspects of scenario exercises (see section 5.2). However, the opportunity to directly participate in the process of scenario development is naturally limited, especially in the case of a broader public on a national or international scale. The way the scenarios are presented then becomes crucial. Effective engagement rather than one-way communication is important; participants need to experience the scenarios in a more interactive format. The related time and resource requirements should not be underestimated. Furthermore, the scenarios must be described in sufficient detail. This does not necessarily include a detailed, rich narrative, although this can be helpful. But it always needs a convincing and illustrative representation of the main strands of development (or scenario mega-trends) and their causes.

Appropriateness of scenario development process

In order to initialize collaborative learning processes and discussions about long-term trends and the consequences of trend development, scenarios first and foremost need to be salient—that is, they must be relevant enough to engage the key target audience in a discussion about either alternative or desirable futures. It is also important that the scenarios are perceived to be as legitimate and credible, but if they fail to present an interesting story they will not succeed in reaching their intended audiences. It is therefore important to ensure that scenarios present appealing, thought-provoking, compelling yet coherent storylines that relate back to a clear focal question. This is especially the case for normative scenario exercises that aim to trigger a wider discussion about desirable futures. Such processes needed to involve a larger number of stakeholders to provide convincing illustration that the scenarios present a wider array of societal opinions (see also section 5.2.5 on involving stakeholders).

Suitability of scenario type and scenario format

Generally speaking, long-term contrasting scenarios have a greater potential to challenge and inspire the mind-sets of policy makers, stakeholders, and the general public than projections or reference scenarios do. Also, in most contexts it seems to be easier to capture the imagination of audiences by presenting compelling qualitative information than detailed quantitative analysis.

Exploratory scenarios that cover a broader framework of social, technological, economic, environmental, and political driving forces in a consistent manner are more suitable to challenge "mental maps" and to challenge participants to think about "weak signals" or "early warnings" of change. If the focus of discussion is on desirable or necessary actions—as opposed to feasible or most likely actions—then normative scenarios seem to be better suited than exploratory ones, as the latter tend to direct participants toward focusing on feasible action.

Match between type of context and communication strategy

In order to be relevant in the context of education and information, scenarios need to be presented in an appealing, easy to understand, and communicative format. The presentation also needs to reach a wide variety of backgrounds and interests. In the past, various formats have been used in diverse outreach processes: the storylines have been presented in form of newspaper articles (see, for example, Kahane 1992), fictive letters from the future (see, for example, NIC 2004), short video teasers (see, for example, EEA 2007), and interactive Web games or other forms of visualization (see also Box 5.13). It is not advisable to operate a broader outreach action process on the basis of a traditional project report only. Having said this, it is important not to underestimate the time and resources it takes to organize a successful outreach information and education process. And—once again—the process is as important as the product.

Outreach activities can be structured along the categories of type of audience and type of discussion desired. The type of audience can be rather broad and include representatives from different ministries and agencies, research, business, and nongovernmental organizations or it can be rather narrow and consist of a specific group of actors from one sector only. Depending on the structure of the target audience, the main questions of interest and the levels of knowledge vary considerably. Furthermore, the broader the target audience is, the more diverse—and potentially contradicting and controversial—the belief systems, policy goals, and institutional routines tend to become. Accordingly, there needs to be more time for broader discussion, which can easily end in a stalemate and thus benefit from being guided by experienced facilitation (see section 5.2.4).

Box 5.13. The PRELUDE 2 action scenario outreach process

The European Environment Agency organized specific outreach workshops to familiarize different audiences with the concept of scenario development and discuss the specific implications of the PRELUDE land use scenarios for Europe. The workshops lasted a full day or two days. Participants were split into working groups and led through presentations of each of the scenarios in a "scenario circuit."

In this circuit, each scenario was presented in a different room, and discussions were moderated by external facilitators. Presentations were supported by video-animations that highlighted key characteristics of the scenarios in an attractive way. Participants discussed and voted on all scenarios on simple radar charts (high, medium, low) with regard to three main criteria: relevance, probability, and desirability. After a group had voted, the results of votes in other groups were shown.

This approach provided a very effective, convincing illustration that individual scenarios can be perceived in very different ways. Comments for the same scenario often ranged from strong refusal ("this is not possible") to strong agreement ("this is already happening"), and the contradictions of these comments triggered learning processes. The skepticism that participants initially expressed about the overall approach seemed to decrease with the number of scenarios "visited" as well as with the time spent exploring these alternative futures (Volkery et al. 2008).

Whether to aim for a broader or narrower audience depends on the overall context conditions and the related main aim—that is, whether the aim is agenda setting for an emerging issue or challenging mind-sets with regard to an established issue. The first case calls for a broader outreach process. The latter case can benefit from mixing diverse opinions and benefits (knowledge cross-fertilization), but if it is necessary to gain confidence it can also be helpful to have the first round of scenario discussions with a restricted target audience only. In all cases it is important to map the overall policy context sufficiently: key policy events, including an analysis of key questions arising in that context, need to be effectively linked into the scenario presentation process.

Fit of scale

If scenarios are to be effective vis-à-vis their educational function, the "fit of scale" is not the most important factor. Since the main aim is to engage a wider range of stakeholders or the public in discussions about alternative futures and their uncertainties, it is more important to have compelling and consistent storylines that help thinking through "what if" questions.

A "fit of scale" factor is more relevant in the context of the informational function of scenarios—that is, if, for example, local communities should be informed about the consequences of nonaction or should be provided with a common platform for discussing key choices, risks, and opportunities for maintaining relevant ecosystem services.

5.5.4 How to use scenarios for decision support and strategic planning

Scenarios can support decision-making processes by providing an analytical framework for finding the suitable, or robust, options with regard to a specific policy target. Scenarios can function as "test beds" (see Wilson 2000):

- Sensitivity/risk assessments evaluate a specific strategic decision, when the need for the decision is known beforehand. Using a series of descriptive and judgmental steps, the process ends with a "go" or "no-go" decision. The series implies identifying future conditions that would be needed to justify a "go" decision, assessing these conditions in each scenario, analyzing how successful the "go" decision would be in each scenario, and finally considering the need to "hedge" or modify the decision to increase its robustness. This approach requires a very clear decision focus, and it is easier to implement in the corporate than the public policy world.
- Strategy evaluations function similarly, but here the scenarios are used as "test beds" to evaluate the viability of a whole existing strategy or policy (robustness testing). The strategy or policy is disaggregated into single thrusts (for example "focus on increasing public awareness," "focus on improving private liability schemes," or "focus on better monitoring"), which are checked for their relevance and likely successes under the different conditions of the scenarios, looking for opportunities that the current strategy or policy addresses or misses and for threats and risks that have not been taken into account. Comparative analysis should allow common success and failure conditions of the strategy to

be identified and henceforth the need for a strategy or policy change but also the need for contingency planning.

However, the complexities and uncertainties of future development might require the development of completely new policies and strategies or a complete organizational overhaul. Scenarios allow the rehearsal of new policies or strategies, analyzing related risks and trade-offs and setting priorities for action. This practice intends to surface long-term blind spots in any policy or strategy and to trigger related learning and decision-making processes (GBN 2006). Strategy development can be done with or without a planning focus (Wilson 2000):

- In strategy development with a planning focus, one scenario is chosen—after analyzing all of them—as a "planning focus" scenario. This is normally the one regarded as most probable by high-level decision makers. It provides the basis for formulating a new strategy or policy, which is subsequently tested for resilience against the other scenarios. Scenario comparison is used to determine the need for modifying the strategy, "hedging," and contingency planning. Although the limitations of a "planning focus" approach are obvious, this approach can help take first steps toward scenario planning in an environment that is more rooted in traditional planning.
- In strategy development without a planning focus, each scenario is analyzed for strategic opportunities and threats that arise, which gives a maximum feasible range of choices. Comparison across the scenarios helps to identify the strategic options that are most robust—that perform reasonably well in any of the scenarios and work best in comparison to all other options considered (see Groves and Lempert (2007) for a detailed description of a quantitative approach to support such analyses).

Scenario exercises are thus effective tools to support better decision making for an uncertain future. The recent literature stresses the process dimension of this endeavor. However, scenario planning is not only a tool to support policy planning, it can also trigger new ways of thinking in policy and organizational development. Scenario planning can be run as singular exercises, which make better sense for existing strategy processes. But it adds real value to decision support if it is institutionalized on a more permanent basis through the creation of a level playing field for a "strategic conversation" among key actors. This implies moving from a more ad-hoc style of developing scenarios toward a permanent process of tracking change in the external environment (van der Heijden 2004).

Furthermore, applying scenarios to decision support and strategic planning is not simply an add-on activity to a scenario exercise. Instead, it needs to be well planned and well staffed—and preferably already be planned alongside the scenario exercise itself. There needs to be a genuine interest in the organization commissioning or undertaking the scenario work in exploring uncertainties in a systematic way, as well as the necessary buy in from high-level management (see discussion on authorizing environments in section 5.2.2). Also, it is essential to understand the organizational context and be able to facilitate complex discussions with clear-cut questions. A good process management is thus equally important as, if not more important than, well-crafted analytically rich scenarios.

It is important to pay attention to some other prominent pitfalls when using scenarios in the context of decision making and strategic planning (Parson et al. 2007, Starr and Randall 2007):

- The challenge of action paralysis: While policy makers might very well be aware of the need to think in alternatives due to future uncertainties and welcome the scenarios, they might be overwhelmed by the magnitude of options and abstain from taking any decisions.
- The challenge of action cosmetics: there is a danger that policy makers instrumentalize scenarios for legitimizing action that has already been taken. The scenarios look nice and attractive, and they symbolize that long-term thinking and preparing for the future is being taken serious, but their actual content is weak and they do not really contribute to the decision-making process (van Notten [2005] calls this the "hollow diamond" effect).
- The challenge of action overshoot: scenarios can misguide political decision-making processes if they are used tactically by policy makers, interest groups, or scientists to highlight problems or foster solutions that are not, but that appear to be, relevant and adequate because they are illuminated by a new, fancy-looking decision support tool.

Appropriateness of scenario development process

In order to receive buy in and support from high-level political decision makers and key stakeholders who have not been part of the development process, scenarios need to be perceived of as legitimate but also salient. The question of legitimacy ("who has been involved in developing the scenarios?" and "how have the scenarios been developed and analyzed?") becomes important to policy makers. This is even more the case for contested or emerging topics with a probability of high impacts. In such cases, the process needs to clearly indicate how different actors have been involved.

Furthermore, in a decision-making context scenarios need to show a true added value—that is, they need to be relevant to the issue at hand and provide a sufficient, nonstereotypical variety of scenario sets. Accordingly, the conditions of and reasons for framing the problem, shaping options, and taking decisions need to be spelled out in a transparent manner. In this context, it is probably less relevant whether the scenarios fully comply with standards of good scientific enquiry.

Suitability of scenario type and scenario format

The interplay between format and use of scenarios is rather important in a decision-making context. For instance, scenarios that address in great detail very specific questions (especially in terms of scientific exploration) are not useful for generic decision support. Conversely, scenarios that are rather broad and aim at capturing a multitude of stressors are likely to be too vague in order to have an efficient impact on concrete political decision-making processes. Scenarios useful in the context of decision support should inform about so-called predetermined elements on the one hand (those forces that can be anticipated with certainty as they are already evolving in their early stages). They frame a strategic corridor in which strategy and policy formulation can take place (Schwartz 1991). Meanwhile, scenarios also need to highlight key uncertainties that could change the validity of key assumptions underpinning strategies or policies. They need to be taken into account to find strategies or policies that are robust enough to function across a wider range of plausible futures (Dewar 1993). Useful scenarios need to allow for policy discretion—that is, if the policies are clearly framed and analyzed throughout the different scenarios, there is no opportunity to experiment with different options, and the scenarios might get rejected as prescriptive.

Also, in a decision-making context, it is important to distinguish between decision unit and decision context in a rigorous manner (or at least try to do so)—that is, to separate the actions an individual decision unit is taking from the actions that are out of the immediate reach and frame the wider decision-making context (see section 5.1). Unfortunately, the evaluative scenario literature on this is rather slim and offers so far only limited insight into the effectiveness of different scenario types for decision support. But it is safe to say that exploratory, alternative scenarios are more helpful to support concrete policy formulation. By contrast, normative scenarios are more helpful in the context of agenda setting and issue framing, with a view to defining a long-term vision and illustrating the magnitude of the challenge of moving from the current status to the future desired status.

Match between type of context and communication strategy

As noted earlier, it is crucial to understand who the key target audiences of a scenario exercise are. Government and public administration are not monolithic blocks, and the expectations toward the scenarios can vary much according to the level of decision making. Experts on the working level usually prefer a more technical presentation than high-level decision makers do, and while this seems to be a trivial statement, it is often exactly at this stage where many scenario outreach processes fail. Moreover, it is important to set the expectations right from the beginning and to stress that scenario exercises are tools to support decision making. They do not replace the need to take decisions.

Often, decision makers in government (especially at higher levels) may seem somewhat biased toward forecasts or reference scenarios forecasts and might be overwhelmed by sets of scenarios that seem confusing rather than helpful in taking decisions effectively ("I don't want choices, I want answers!"). Providing a list of examples of failed single-point forecasts in the past and a clear-cut illustration of the way scenarios could be used—rather than presenting a battery of detailed findings—can help overcome these rather cultural barriers (Wilson 2000).

If scenarios are to guide strategic planning, it is helpful to work with a list of clear cut questions that get political decision makers to identify the consequences of action or nonaction and respective opportunities for policy making, especially with a focus on their own organization in each scenario, although these might seem unattractive from today's perspective. It is relevant to determine who should be targeted and also when. Any action should be guided by a mapping of key policy events, or milestones of decision making, and the related decision-making needs. If the respected policy field and its key actors are rather closed and dominated by well-established belief systems, the primary task of using scenarios is then to overcome agency resistance to change, which requires a rather broad, open presentation with built-in persuasion effects. But scenarios might also be used to rehearse strategies on how to break opposition by key societal stakeholders and develop alliances with new stakeholders.

Fit of scale

A fit of scale is highly relevant for scenarios to be useful in decision support and strategic planning. The implications of different strategic decisions must be clearly visible and understandable on the right scale. Otherwise, the scenarios will be regarded as nice, even visionary work that is of little added value to concrete processes of decision making (see Box 5.14).

Box 5.14. Using scenarios "out of context"

Scenarios are not bound to the original thematic context and purpose they have been developed for. Especially under conditions of time or financial resource constraint that do not allow for developing new scenarios, the results from other exercises can be used "out of context." Due to the increasing economic, social, and ecological interlinkages across regions and scales, a set of scenarios that contains a qualitative description of alternative futures based on a set of different, yet coherent assumptions about driving forces normally offer enough opportunities to link the scenarios to their own organizational context and to explore uncertainties of long-term trends.

Using scenarios out of context works best with rather broad thematic scenarios in settings that aim at general discussions—thus especially for public information and learning, but also when more indirect forms of decision support are required, such as, for example, clarifying an issue's importance, shaking up habitual thinking, or stimulating creativity (Parson et al. 2007). Direct forms of decision support, such as informing specific decisions, normally require, on the contrary, tailor-made scenarios that reflect the relevant drivers, uncertainties, and information needs of the organization.

However, if scenarios are indeed used out of context it is necessary to make sure that all the relevant information is available that is needed to fully understand the scenarios. It is not enough to simply take a report from the shelf, as most often this does not offer sufficient information on the choice and meaning of assumptions, details of analysis, and interpretation of results. Furthermore, if information is borrowed from different scenario sets, a key requirement is to ensure a basic compatibility of scenario assumptions to avoid inconsistencies and contradictions. Especially in larger outreach settings, participants are bound to scrutinize the scenarios in detail, and a failure to provide clear answers can easily undermine the legitimacy of the process.

Quick Reference Charts

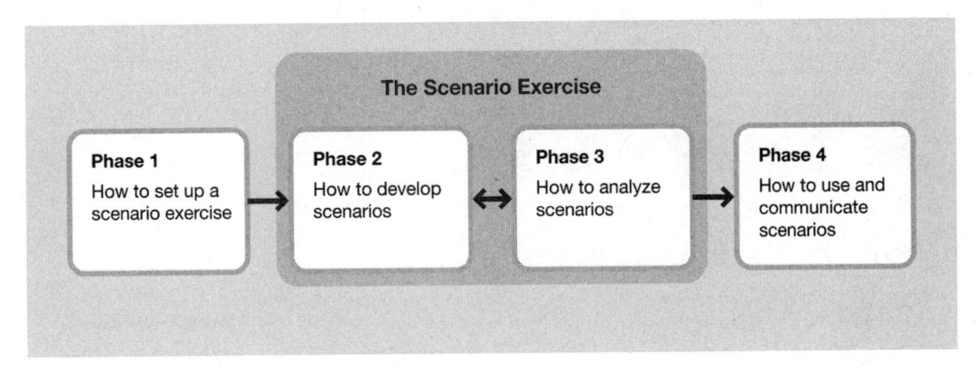

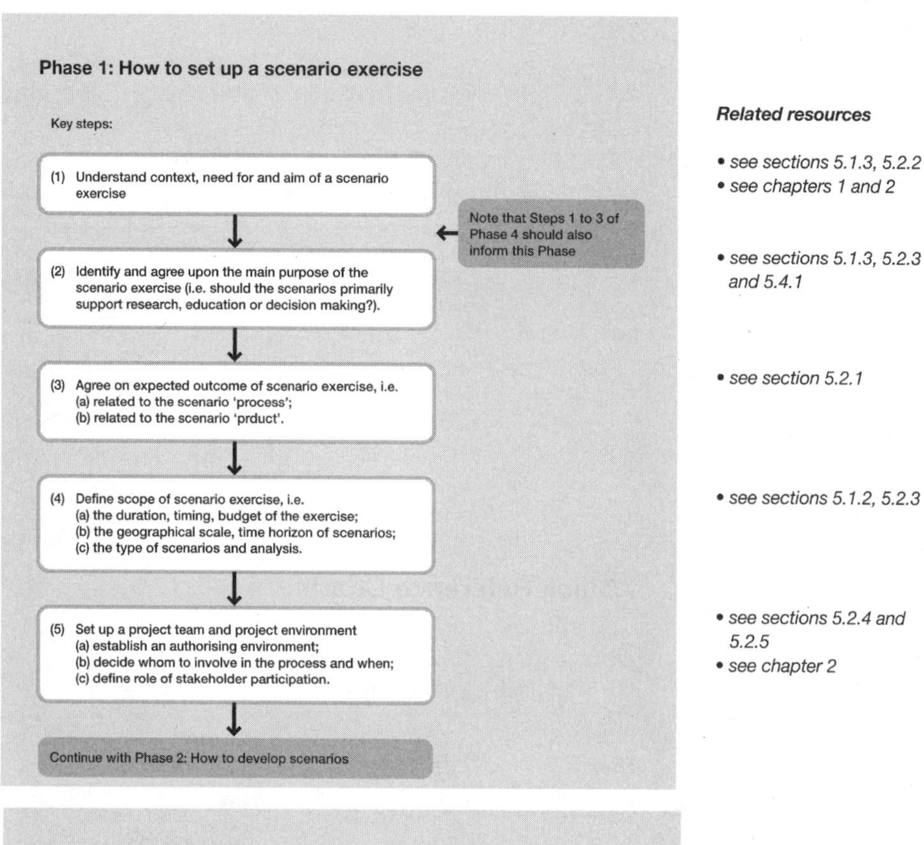

Related resources

- *see sections 5.1.3, 5.2.2*
- *see chapters 1 and 2*
- *see sections 5.1.3, 5.2.3 and 5.4.1*
- *see section 5.2.1*
- *see sections 5.1.2, 5.2.3*
- *see sections 5.2.4 and 5.2.5*
- *see chapter 2*

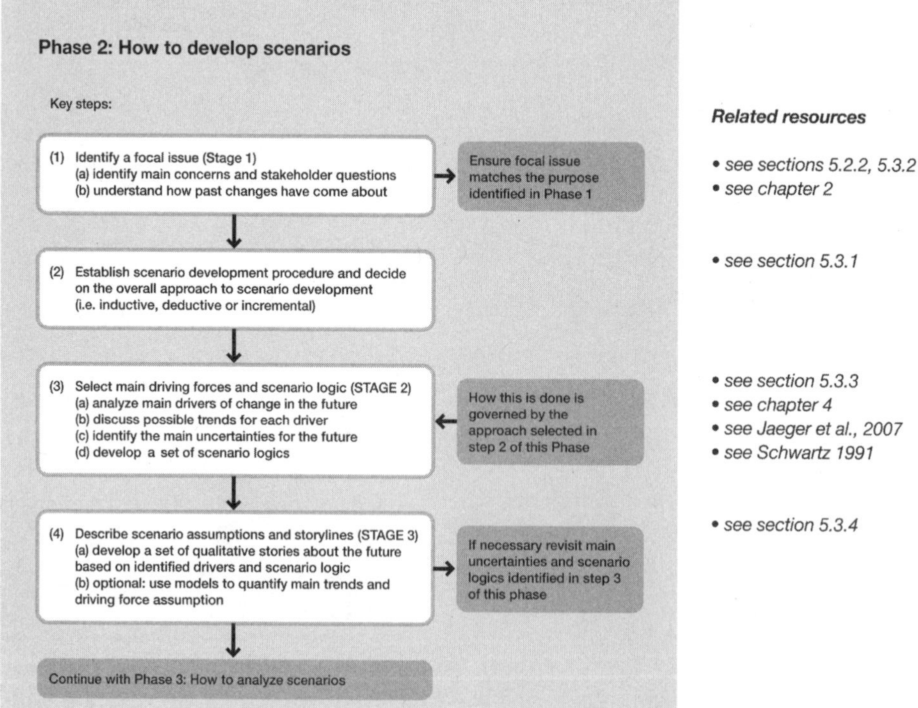

Related resources

- *see sections 5.2.2, 5.3.2*
- *see chapter 2*
- *see section 5.3.1*
- *see section 5.3.3*
- *see chapter 4*
- *see Jaeger et al., 2007*
- *see Schwartz 1991*
- *see section 5.3.4*

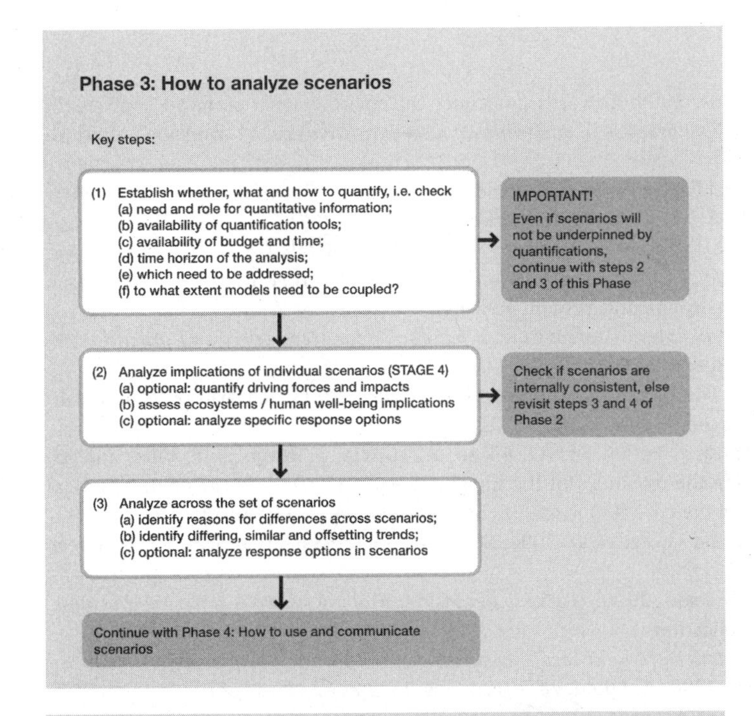

Related resources

- *see sections 5.1.3, 5.2.2 (whether)*
- *see sections 5.4.1, 5.4.2, 5.4.4 (what and how)*
- *see EEA, 2001*

- *see section 5.4.2*
- *see MA 2005a*
- *see UNEP 2002*

- *see section 5.4.3*
- *see MA 2005a*
- *see UNEP 2002*

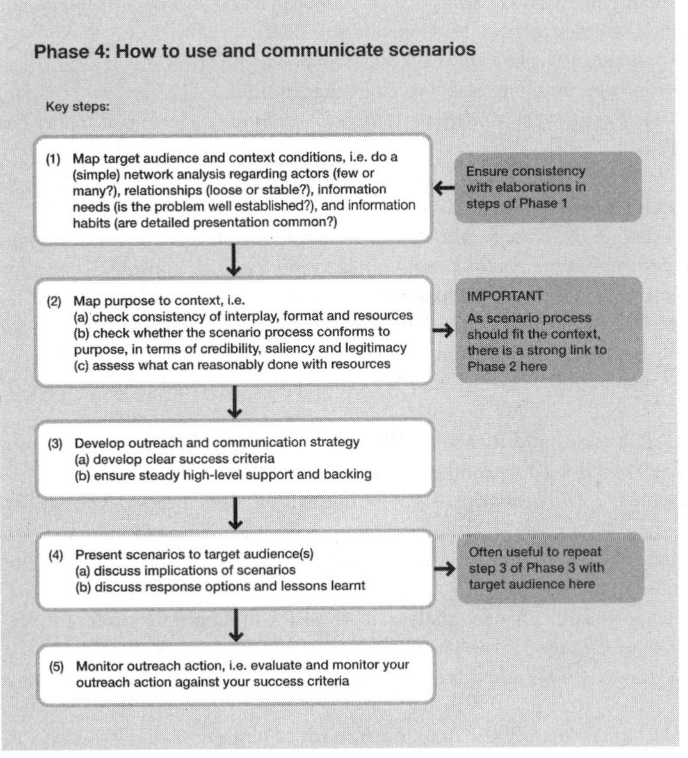

Related resources

- *see sections 5.1.3, 5.2.2 5.5.1*
- *see chapters 1 and 2*

- *see sections 5.2.1, 5.5.2, 5.5.3, 5.5.4*

- *see sections 5.5.2, 5.5.3, 5.5.4*

- *see chapters 1 and 2*

- *see chapters 1 and 2*

References

Alcamo, J., and T. Henrichs. 2008. Towards guidelines for environmental scenario analysis. In *Environmental futures: The practice of environmental scenario analysis*, ed. J. Alcamo. Oxford: Elsevier.

Alcamo, J., R. Leemans, and E. Kreileman, eds. 1998. *Global change scenarios of the 21st century: Results from the IMAGE 2.1 model*. Oxford: Elsevier.

Alcamo, J., T. Henrichs, and T. Rösch. 2000. *World water in 2025—Global modelling and scenario analysis for the World Water Commission on Water for the 21st Century*. Report A0002, Kassel, Germany: Center for Environmental Systems Research, University of Kassel.

Armstrong, J., ed. 2001. *Principles of forecasting: a handbook for researchers and practitioners*. Norwell, MA: Kluwer Academic Publishers.

Börjeson, L., M. Höjer, K. Dreborg, T. Ekvall, and G. Finnveden. 2006. Scenario types and techniques: Towards a users guide. *Futures* 38 (7): 723–39.

Bouwmans, R., R. Costanza, J. Farley, M. A. Wilson, R. Portela, J. Rotmans, F. Villa, and M. Grasso. 2002. Modeling the dynamics of the integrated earth system and the value of global ecosystem services using the GUMBO model. *Ecological Economics* 41:529–60.

Carpenter, S., E. Bennett, and G. Peterson. 2006. Scenarios for ecosystem services: An overview. *Ecology and Society* 11 (1): 29.

Cosgrove, W., and F. Rijsberman. 2000. *World water vision: Making water everybody's business*. London: Earthscan Publications.

Dewar, J. 1993. *The importance of "wild card" scenarios*. Santa Monica, CA: RAND.

Ducot, C., and H. Lubben. 1980. A typology for scenarios. *Futures* 12 (1): 15–57.

EEA (European Environment Agency). 2001. *Scenarios as tools for international environmental assessments*. Environmental issues series No 24. Copenhagen: EEA.

EEA. 2005. *European environment outlook*. Report 2/2005. Copenhagen: EEA.

EEA. 2006. *Urban sprawl in Europe*. Briefing 04-2006. Copenhagen: EEA.

EEA. 2007. *Land use scenarios for Europe: Modelling at the European scale*. Technical report No 09/2007. Copenhagen: EEA.

Eickhout, B., H. van Meijl, A. Tabeau, and T. van Rheenen. 2007. Economic and ecological consequences of four European land use scenarios. *Land Use Policy* 24 (3): 562–75.

Evans, K., S. J. Velarde, R. Prieto, S. N. Rao, S. Sertzen, K. Davila, P. Cronkleton, and W. de Jong. 2006. *Field guide to the future: Four ways for communities to think ahead*. Nairobi: Center for International Forestry Research, ASB, and World Agroforestry Centre.

FAO (Food and Agriculture Organization). 2002. *World agriculture: Towards 2015/2030—An FAO perspective*. London: Earthscan Publications.

Gallopin, G., and F. Rijsberman. 2000. Three global water scenarios. *International Journal of Water* 1 (1): 16–40.

Gallopin, G., A. Hammond, P. Raskin, and R. Swart. 1997. *Branch points: global scenarios and human choice*. Stockholm: Stockholm Environment Institute.

GBN (Global Business Network). 2006. *Scenarios, strategy, and action. Guide & workbook*. San Francisco: GBN.

Gordon, T., and H. Hayward. 1968. Initial experiments with the cross-impact matrix method of forecasting, *Futures* 1:100–16.

Groves, D. G., and R. J. Lempert. 2007. A new analytic method for finding policy-relevant scenarios. *Global Environmental Change* 17:78–85.

Grübler, A., and N. Nakicenovic. 2001. Identifying dangers in an uncertain climate. *Nature* 412:15.

Grübler, A., B. O'Neill, and D. van Vuuren. 2006. Avoiding hazards of best-guess climate scenarios. *Nature* 440 (7085):740.

Helmer, O. 1983. *Looking forward: A guide to futures research*. London: SAGE Publications.

Hughes, B. 1999. *International futures: choices in the face of uncertainty*. Oxford: Westview Press.

IEA (International Energy Agency). 2007. *World energy outlook*. Paris: IEA.

IPCC (Intergovernmental Panel on Climate Change). 1992. *Climate change 1992: Supplementary report to the IPCC Scientific Assessment*. Cambridge, U.K.: Cambridge University Press.

IPCC. 2000. *Special report on emission scenarios*. Cambridge, U.K.: Cambridge University Press.

IPCC. 2001. *Climate change 2001—Synthesis report*. Cambridge, U.K.: Cambridge University Press.

IPCC. 2007. *Climate change 2007—IPCC fourth assessment report*. Cambridge, U.K.: Cambridge University Press.

Jaeger, J., D. Rothman, C. Anastasi, S. Kartha, and P. van Notten. 2007. Training Module 6: Scenario development and analysis. In *GEO resource book: A training manual on integrated environmental assessment and reporting*. Nairobi and Winnipeg, Canada: U.N. Environment Programme and International Institute for Sustainable Development.

Jungk, R., and N. Müllert. 1987. *Future workshops: How to create desirable rutures*. London: Institute for Social Inventions.

Kahane, A., ed. 1992. The Mont Fleur scenarios—What will South Africa be like in 2002? *GBN Deeper News* 7(1). Emeryville, CA: Global Business Network.

Kahn, H. 1960. *On thermonuclear war*. Oxford: Oxford University Press.

Kahn, H., and A. Wiener. 1967. *The year 2000*. New York: Macmillan.

Kok, K., R. Biggs, and M. Zurek. 2007. Methods for developing multiscale participatory scenarios: Insights from Southern Africa and Europe. *Ecology and Society* 13 (1): 8.

Leemhuis, J. 1985. Using scenarios to develop strategies. *Long Range Planning*, 18 (2): 30–37.

Lempert, R., N. Nakicenovic, D. Sarewitz, and M. Schlesinger. 2004. Characterizing climate-change uncertainties for decision makers. *Climatic Change* 65: 1–9.

MA (Millennium Ecosystem Assessment). 2005a. *Ecosystems and human well-being: Scenarios*. Washington DC: Island Press.

MA. 2005b. *Ecosystems and human well-being: Multiscale assessments*. Washington DC: Island Press.

Meadows, D. H., D. L. Meadows, J. Randers, and W. W. Behrens 1972. *The limits to growth*. New York: Universe Books.

Mesarovic, M. and E. Pestel. 1974. *Mankind at the turning point*. New York: Dutton.

Mitchell, R., W. Clark, D. W. Cash, and N. Dickinson, eds. 2006. *Global environmental assessments: Information, institutions, and influence*. Cambridge, MA: The MIT Press.

NIC (National Intelligence Council). 2004. *Mapping the global future*. Report of the National Intelligence Council's 2020 project. Washington, DC: U.S. Government Printing Office.

OECD (Organisation of Economic Co-operation and Development). 2008. *OECD's environmental outlook to 2030*. Paris: OECD.

Ogilvy, J., and P. Schwartz. 1998. Plotting your scenarios. In *Learning from the Future*, ed. L. Fahey and R. Randell. New York: John Wiley & Sons.

Parson, E., V. Burkett, K. Fisher-Vanden, D. Keith, L. Mearns, H. Pitcher, C. Rosenzweig, and M. Webster. 2007. *Global-change scenarios: Their development and use*. Report for the U.S. Climate Change Science Program and the Subcommittee on Global Change Research.

Pereira, E., C. Queiroz, H. Pereira, and L. Vicente. 2005. Ecosystem services and human well-being: A participatory study in a mountain community in Portugal. *Ecology and Society* 10 (2): 14.

Pielke, R., T. Wigley, and C. Green. 2008. Dangerous assumptions. *Nature* 452: 531–32.

Raskin, P. 2005. Global scenarios: Background review for the Millennium Ecosystem Assessment. *Ecosystems* 8 (2): 133–42.

Raskin, P., G. Gallopin, P. Gutman, A. Hammond, and R.Swart. 1998. *Bending the curve: Towards global sustainability*. Stockholm: Stockholm Environment Institute.

Raskin, P., T. Banuri, G. Gallopin, P. Gutman, A. Hammond, R. Kates, and R. Swart, R. 2002. *Great transitions: The promise and lure of times ahead*. Boston: SEI-B / Tellus Institute.

Reidsma, P., T. Tekelenburg, M. van den Berg, and R. Alkemade. 2006. Impacts of land use change on biodiversity: An assessment of agricultural biodiversity in the European Union. *Agriculture, Ecosystems and Environment* 114:86–102.

Richels, R. G., A. S. Manne, and T. M. L. Wigley. 2004. *Moving beyond concentrations: the challange of limiting temperature change.* AEI–Brooking Joint Center for regulatory studies. Washington DC: Brookings Institution.

Rienks, W. A. 2008. The future of rural Europe. An anthology based on the results of the Eururalis 2.0 scenario study. Wageningen, The Netherlands., Wageningen University Research and Netherlands Environmental Assessment Agency.

Ringland, G. 2002. *Scenarios in public policy.* Chichester, U.K.: John Wiley & Sons.

Rothman, D. 2008. Environmental scenarios—A survey. In *Environmental futures: The practice of environmental scenario analysis.* Ed. J. Alcamo. Oxford: Elsevier.

Rotmans, J., and B. de Vries, eds. 1997. *Perspectives on global futures: the TARGETS approach.* Cambridge, U.K.: Cambridge University Press.

Rounsevell, M. D. A., F. Ewert, I. Reginster, R. Leemans, and T. R. Carter. 2005. Future scenarios of European agriculture land use. II. Projecting changes in cropland and grassland. *Agriculture, Ecosystems & Environment* 107:117–35.

Scearce, D., K. Fulton, and the Global Business Network community. 2004. *What if? The art of scenario thinking for nonprofits.* Emeryville, CA: Global Business Network.

Schneider, S. H. 2001. What is "Dangerous" Climate Change? *Nature* 411:17–19.

Schneider, S. H. 2002. Can we estimate the likelihood of climatic changes at 2100? An editorial comment. *Climatic Change* 52:441–51.

Schwartz, P. 1991. *The Art of the long view.* New York: Currency Doubleday.

Starr, J., and D. Randall. 2007. Growth scenarios: Tools to resolve leaders' denial and paralysis. *Strategy and Leadership* 35 (2): 56–59.

Stewart, J. 1993. Future state visioning—A powerful leadership process. *Long Range Planning* 26 (6): 89–98.

Toth, F., E. Hizsnyik, and W. Clark. 1989. *Scenarios of socioeconomic development for studies of global environmental change: A critical review.* IIASA Report RR-89-4. Laxenburg, Austria: International Institute of Applied Systems Analysis.

UNEP (United Nations Environment Programme). 2002. *Global environment outlook 3.* London: Earthscan Publications.

UNEP. 2007. *Global environment outlook 4.* London: Earthscan Publications.

van der Heijden, K. 2004. *Scenarios: the art of strategic conversation.* New York: John Wiley & Sons.

van Notten, P. 2005. *Writing on the wall: Scenario development in times of discontinuity*, Dissertation.com, Florida.

van Notten, P., J. Rotmans, M. van Asselt, and D. Rothman. 2003. An updated scenario typology. *Futures* 35:423–43.

van Vuuren, D., M. den Elzen, P. Lucas, B. Eickhout, B. Strengers, B. van Ruijven, S. Wonink, and R. van Houdt. 2007. Stabilizing greenhouse gas concentrations at low levels: An assessment of reduction strategies and costs. *Climatic Change*, 81:119–59.

van Vuuren, D., et al., 2008. Forthcoming.

Volkery, A., T. Henrichs, Y. Hoogeveen, and T. Ribeiro, T. 2008. Your vision or my model? Lessons from participatory land use scenario development on a European scale. *Systemic Action & Practice Research*, forthcoming.

Wack, P. 1985. The gentle art of reperceiving. *Harvard Business Review* (article in two parts). September–October:73–89; November–December:139–50.

WBCSD (World Business Council for Sustainable Development). 1997. *Exploring sustainable development: WBCSD global scenarios 2000–2050 summary brochure.* Geneva: WBCSD.

Webster, M. D., M. H. Babiker, M. Mayer, J. M. Reilly, J. Harnisch, R. Hyman, M. C. Sarofim, and C. Wang, C. 2002. Uncertainty in emissions projections for climate models. *Atmospheric Environment.* 36 (22): 3659–70.

Webster, M. D., C. Forest, J. M.Reilly, M. Babiker, D. Kicklighter, M. Mayer, R. Prinn et al. 2003. Uncertainty analysis of climate change and policy response. *Climatic Change* 61:295–320.

Westhoek, H., M. van der Berg, and J. Bakkes. 2006. Scenario development to explore the future of Europe's rural areas. *Agriculture, Ecosystems & Environment* 114 (1): 7–20.

Wilkinson, A., and E. Eidinow. 2008. Evolving practices in environmental scenarios: a new scenario typology. *Environmental Research Letters* 3.

Wilson, I. 2000. From scenario thinking to strategic action. *Technological Forecasting and Social Change* 65 (1): 23–29.

Wollenberg, E., D. Edmunds, and L. Buck. 2000. *Anticipating change: Scenarios as a tool for adaptive forest management: A guide*. Bogor, Indonesia: Center for International Forestry Research.

WRI (World Resources Institute). 2008. *Ecosystem services—A guide for decision makers*. Washington, D.C.: WRI.

Zurek, M., and T. Henrichs. 2007. Linking scenarios across geographical scales in international environmental assessments. *Technological Forecasting and Social Change* 74 (8): 1282–95.

6 Assessing Intervention Strategies

R. David Simpson and Bhaskar Vira

What is this chapter about?

This chapter looks at the effectiveness of strategies that respond to the degradation of ecosystems that provide services to society. First we consider who might develop strategies for responding to ecological degradation and what considerations and constraints might shape the strategies they can choose. The choice among strategies inevitably involves trade-offs between competing objectives. The chapter looks at some techniques practitioners may use in evaluating such trade-offs. Then we consider several strategies that have already been used to address the degradation of ecosystems. This discussion is intended to assist practitioners who are considering different types of intervention options by highlighting the advantages as well as the pitfalls of various alternatives. The final main section discusses how the results of assessment procedures may be used to inform decision-making processes, as well as how the attributes of decision-making processes may affect the motivations for, and uses of, assessment procedures.

6.1 How to choose from a menu of possible strategies

Section's take-home messages

- It is important to ensure congruence between the diagnosis of a problem, the capacity of agents to respond to the problem, and the scale at which an intervention is implemented.
- For interventions to be effective, knowledge about the problem is fundamental, but it needs to be supported by appropriate institutions and social contexts to ensure that instrumental interventions achieve their desired objectives.
- If knowledge is incomplete, institutions are imperfect, or social conditions are inappropriate, practitioners may not be in a position to implement desirable strategies without addressing these fundamental constraints.
- The five arenas of action that actors can potentially be involved in are provision of knowledge, reform of institutions and governance, societal and behavioral innovation, use of markets and incentives, and development of improved technologies.

- Identification of constraints helps to focus attention on the enabling conditions that are necessary for an intervention to work.

In deciding on the best response, if any, to ecological degradation at a particular time and place, several interrelated questions must be addressed:

- What is the ecosystem change/loss in human well-being that needs to be addressed, and why?
- Who will respond?
- Which strategies will they choose?
- How will these strategies be structured?
- What will their effects be on both ecosystems and human well-being?

These key questions are addressed in this chapter, starting in this section with the first three questions.

6.1.1 How to determine who can and should respond

Whether one wishes to evaluate the success of a measure that has been implemented in the past or the likely efficacy of one that is now being contemplated, a critical question is "Does this strategy propose an effective measure for dealing with the problem at hand?" To answer this question, one must first understand what the problem *is*. For our purposes, the important point is to highlight that problem identification and definition are usually the first stage of a planned intervention or evaluation strategy. Furthermore, decision makers usually frame problems based on their own perceptions, knowledge, and understanding of the drivers of change (Adams et al. 2003). It is possible that different decision makers will focus on different elements of a perceived empirical reality; this framing of the problem is also likely to inform the range of solutions that are explored.

Strategies that respond to ecological degradation and the consequent loss of human well-being will be futile if those planning them do not have the capability to implement them. In this sense, "capability" may refer to the understanding needed to address a complex problem, the political, organizational, or economic power to overcome resistance, or a combination of the two. A good place to begin to consider these matters is with a discussion of "scale."

Local events can have global implications. Even when physical effects cannot be demonstrated to occur over great distances or from one time to another, local actions can still have consequences for the global community. Someone may derive no material benefit from the continuing survival of rhinoceroses in Africa or polar bears in the Arctic, for example, but may still care enough about preventing their extinction to be willing to contribute to their survival. Equally, a number of processes that occur at local scales may have profound local implications, but they need not necessarily cascade to other scales of impact; localized contamination of lands due to industrial effluent may not necessarily affect a very large area. Such local issues are in no way less important, but the level at which the problem is perceived, and the strategies to respond to the problem, are likely to be operating at a very different scale to an issue that commands regional, national, or global attention. Hence, the question of scale (of impact and response) is critical.

While local actions may have global implications, altering local incentives is sometimes sufficient to solve problems of global importance. Consider, for example, overfishing. While the depletion of fisheries may have global implications for ecosystem function, an effective intervention may in some instances involve only convincing local fishers that it is in their interest to conserve the resource on which they depend.

Thus, in evaluating the effectiveness of strategies for conservation and preservation of ecosystem services, a critical issue is whether such strategies have been implemented on a spatial and temporal scale sufficiently broad as to lead to effective action. This is a matter not only of geography and time but of political boundaries and knowledge as well.

Bohensky and Lynam (2005) provide a useful framework for thinking about these issues, which we will borrow here. They note that effective strategies arise when impact scope, awareness, and power converge. Impact scope is the spatial and temporal scale at which the consequences of ecosystem change are felt. It is necessarily a matter of degree and interpretation, and the scope of analysis may vary depending on the aspect of the problem being considered. Awareness is simply knowing that ecosystem change is occurring and driving changes in well-being. Different actors may cognitively frame these changes very differently, thereby shaping their perceptions of the problem. Power is the ability to effect change. "Power" means that affected people are able to either address the problem themselves or to provide sufficient incentives to others to change their behavior.

It is easy to identify situations in which impact scope, awareness, and power overlap as well as others in which they do not. Timber concessions and oil palm plantations in the tropics, for example, often result in deforestation in areas in which local people depend on ecosystem services such as water purification, erosion protection, and habitat for harvested wildlife. Local people are often aware of the benefits they obtain from existing forests, but they may lack the power to prevent land clearing. If urban or international interests stand to benefit from deforestation, and these interests are politically more powerful than community-based or indigenous groups, no effective response may be possible.

These possibilities are illustrated in Figure 6.1. The top figure illustrates the situation in which a local community is aware of the effects of deforestation but does not have the power to resist it. In the bottom figure we envisage a situation in which two things have happened. First, the power of the local community has increased, as represented by an increase in the size of the most darkly shaded ellipse. Second, the local community has acquired a greater ability to affect outcomes, as represented by a shift in the most darkly shaded ellipse to overlap the "awareness" and "impact scope" ellipses more. This may have resulted from both an increase in its overall authority and the transfer to local groups of some authority that formerly lay with external actors, for instance through the introduction of decentralized and devolved forms of forest governance.

This way of thinking highlights the need to have the knowledge, institutional, and social prerequisites for effective strategies in place and to ensure that these converge in terms of impact scope, awareness, and power. The next question concerns identifying which strategies are most appropriate and the feasibility of the chosen options.

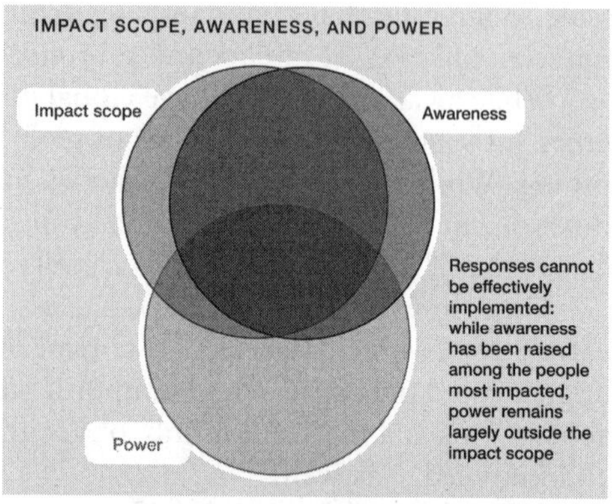

Figure 6.1a. Impact scope, awareness, and power.
Source: Adapted from Bohensky and Lynam (2004)

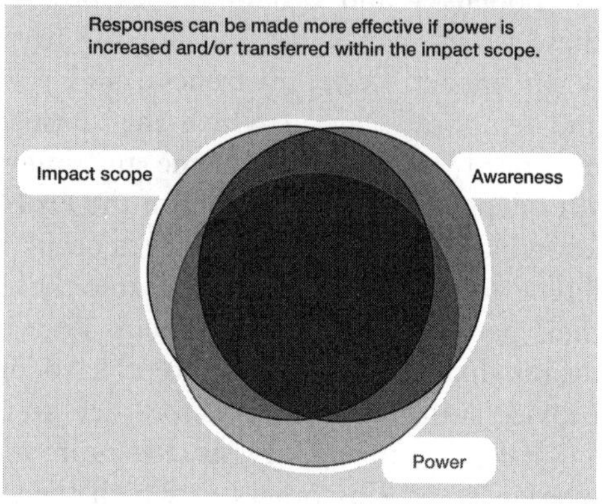

Figure 6.1b. Responses can be made more effective if power is increased and/or transferred within the impact scope.
Source: Adapted from Bohensky and Lynam (2004)

6.1.2 How to decide which strategies are appropriate

The Millennium Ecosystem Assessment (MA) identifies five types of problems that might result in the degradation of ecosystems and motivate strategies (MA 2005a):

- Inappropriate institutional and governance arrangements, including the presence of corruption and weak systems of regulation and accountability;
- Market failures and the misalignment of economic incentives;
- Social and behavioral factors, including the lack of political and economic power of some groups (such as poor people, women, and indigenous peoples) that are particularly dependent on ecosystem services or harmed by their degradation;
- Underinvestment in the development and diffusion of technologies that could increase the efficient use of ecosystem services and could reduce the harmful impacts of various drivers of ecosystem change; and
- Insufficient knowledge (as well as the poor use of existing knowledge) concerning ecosystem services and management, policy, technological, behavioral, and institutional strategies that could enhance benefits from these services while conserving resources.

With such a wide range of problems, it is only to be expected that deciding when and where to apply pressure for strategies will sometimes be difficult. In general, the choices of who should respond and how depend on context and the capabilities of actors.

To see which actors might best be involved in which contexts, consider the five arenas of actions suggested. Summarizing briefly, they are provision of knowledge, reform of institutions and governance, societal and behavioral innovation, the use of markets and incentives, and the development of improved technologies.

Figure 6.2 presents a schematic representation of these potential actions. Focus for now on the left side of the figure. The far-left column presents levels or "tiers" of potential actions. The first tier is fundamental. An adequate knowledge base for action is a prerequisite for taking any meaningful action. Without sufficient knowledge of biological and social facts it would be impossible to structure effective strategies. Below fundamental knowledge, two categories of actions are listed that will, in turn, enable more instrumental strategies. These are the categories labeled "institutions and governance" and "society and behavior." We characterize the function of these strategies as "enablement," as actions in these realms establish the preconditions for enacting instrumental strategies. This last category, "instruments" describes specific actions taken to achieve particular ends, which include the deployment of technological interventions and/or the use of markets and incentives.

To give some examples, an instrumental (third tier) response might be to pass regulations intended to restrict deforestation. Such regulations would not be effective without second-tier prerequisites, such as a legal system capable of enforcing them, or a "culture of compliance" in which citizens generally accept the decisions of their government as fair and just—and abide by them. Another instrumental response to ecosystem change might be to subsidize research on more benign agricultural technologies. Again, such subsidies would not be effective absent certain social and institutional preconditions, such as the existence of a university system embodying the required scientific expertise or social norms under which the best qualified individuals can pursue careers in research, regardless of background, economic status, ethnicity, or gender.

Figure 6.2 presents a sort of cascade, in which knowledge creates the preconditions for institutional and social strategies, which in turn create the preconditions for specific instrumental strategies. Thus a sequential procedure would involve identifying whether there is appropriate foundational knowledge, checking whether the appropriate enabling conditions are in place, and then identifying if relevant and appropriate instruments can be deployed.

There are, however, feedbacks. For example, technological improvements can lead to the generation of knowledge that will guide other strategies, as well as for the reform of institutions and social relations. We will suppose, however, that the foundational and enabling prerequisites must be in place if instrumental strategies are to be effective.

Note that the far right column in Figure 6.2 is labeled "initiating actors," to distinguish between actors who might participate in particular approaches or be affected by them and those who would typically initiate them. This is in keeping with the general idea that, in assessing strategies, it is important to know who must be given an incentive to act in order for the strategy to be taken up. That is, who must make the policy choices required to get the action under way?

The identification of initiating actors is based on political and economic considerations. The essential idea here is that each type of action represents a kind of public good: something that, when provided by one actor, is available to many. Knowledge has this character: when knowledge is in the public domain, it is available for the

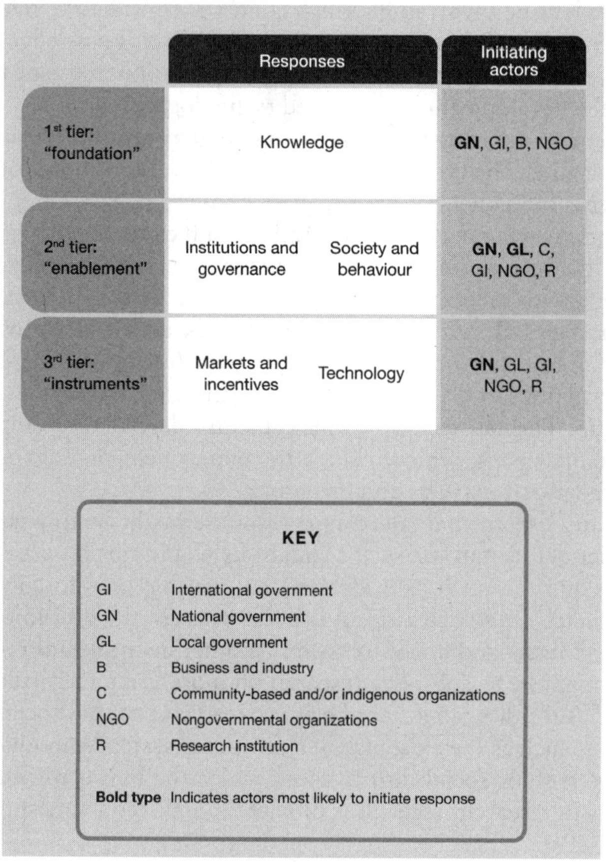

	Responses		Initiating actors
1st tier: "foundation"	Knowledge		**GN**, GI, B, NGO
2nd tier: "enablement"	Institutions and governance	Society and behaviour	**GN**, **GL**, C, GI, NGO, R
3rd tier: "instruments"	Markets and incentives	Technology	**GN**, GL, GI, NGO, R

KEY

GI	International government
GN	National government
GL	Local government
B	Business and industry
C	Community-based and/or indigenous organizations
NGO	Nongovernmental organizations
R	Research institution
Bold type	Indicates actors most likely to initiate response

Figure 6.2. Responses and actors.

benefit of all. A similar statement might be made concerning good governance: reforms that improve the functioning of government provide benefits to all of the governed—that is, to everyone. The ecosystem services of concern here are also generally public goods: enhancements made by one person benefit many. There is, of course, an important caveat here, since while some types of collective goods benefit many, they do allow for the exclusion of certain groups, and this may not always be equitable or fair. We should not assume that all ecosystem services are equally accessible to all groups in society.

Conventional wisdom has it that public goods must be provided by public authorities. No one individual or group has the incentive to provide a public good, as they will bear all the costs but appropriate only some of its benefits (i.e., others will ride free on the private provision of a public good). Therefore, in many instances we identify one or another level of government as the most likely and appropriate locus for initiating an intervention strategy.

Our focus on public entities as initiating agents is not intended to minimize the role of other actors. Once a strategy is set in motion by one type of actor, some or every other type might play useful roles. For example, local governments may enact environmental regulations, but the laws' efficacy depends on compliance among community and business actors. Moreover, private-sector and civil society actors interact with public-sector actors to set the social agenda. It is therefore a simplification to treat public entities as initiating agents, when public priorities are generated by a complex interaction of societal interests. Practitioners may want to enlist business or community constituencies in order to generate support for particular strategies. This section focuses, however, mostly on those actors who can directly initiate strategies.

The acquisition of basic knowledge has been and is likely to continue to be funded most prominently by national governments. While there are compelling arguments to be made that the spillovers of knowledge are global, international cooperation in research financing remains relatively rare. Private industry has also, on occasion, been a major funder of basic research, and that may continue. Nongovernmental organizations (NGOs) and community-based and indigenous organizations may also play a role in this area, although knowledge that is embedded within such organizations is often not more widely used until it has been formalized and published, usually by outsiders (who act as promoters or propagators of this knowledge and in some cases illegitimately appropriate it for their own selfish ends). Research organizations are often very important nodes in the knowledge process, but they are usually responsive to external needs for new products or processes rather than initiators of the information acquisition process. "Blue skies" research, when it occurs, is also usually supported by government or private foundations, and it is rare for institutions to have resources that enable an investment in such fundamental research without external support.

In the second tier, national and local governments and community-based and indigenous organizations are identified as the most likely initiators of reforms in governance and social relations. The need for action can arise at any of several levels: internationally, for example, to address greenhouse gas emissions and global warming; at the level of the nation or region, to address major pollutants or major ecosystem disruption; or at the level of local pollution or habitat issues. While the general conclusion of extensive literature on this issue (see, e.g., Olson 1965, Ostrom 1990, Baland and Platteau 1996) is that effective management is more likely in smaller and more homogeneous groups, incorporating the needs of all affected parties (often referred to as stakeholders) in decision making is a prerequisite for dealing with ecological issues at any level.

In addition to government and community organizations, NGOs have on occasion provided the impetus for institutional and social reforms. Research organizations are also listed as an initiating agent, as the findings of social scientists and other researchers have sometimes spurred action. While international action would be appropriate in many instances, issues of sovereignty have often led international efforts at influencing domestic institutional reform to be less successful or less ambitious.

In the third tier, that of instrumental strategies, action is generally initiated by national or local governments. This is almost a tautology in the case of regulations, which are imposed and enforced by government. International agreements have been pivotal in many strategies, such as the Montreal and Kyoto Protocols and the Convention on Biological Diversity. These international agreements have often contained provisions for particular instruments, such as trading in emissions or credit for reductions under the Clean Development Mechanism of the Kyoto

Protocol. In some instances community-based or indigenous organizations adopt specific instrumental strategies, such as restrictions on harvesting or resource use. NGOs and research organizations have also been active in some instrumental strategies, particularly in developing, promoting, and distributing technology.

The distinction between strategies in the second and third tiers of Figure 6.2 is generating the conditions to enable more specific strategies and adopting specific instruments tailored to particular ecological circumstances. Each response has its merits and drawbacks. The virtue of a more direct response—passing regulation to reduce deforestation, for example, or subsidizing research on sustainable agricultural technologies—is that it focuses on the problem of interest rather than matters that may, from an ecological perspective, be more peripheral. In other words, if an instrumental approach "works," the problem will be "solved."

The potential drawback is that such direct approaches presume certain institutional and social infrastructure. For instance, regulations restricting deforestation presume a competent regulatory authority, fair and consistent enforcement, and private actors who will be dissuaded from illegal action by the threat of sanction. Similarly, subsidies to research presume a research sector capable of producing innovations.

Would it be more effective to respond to wasteful degradation of ecosystem services with institutional or social and behavioral interventions rather than regulatory and technological ones? There are pluses and minuses. On the positive side, such strategies are intended to create the preconditions for the subsequent application of direct strategies. On the negative side, institutional and social strategies may dissipate the effect of resources intended to be devoted to the preservation of imperiled ecosystems. Consider, for example, a couple of strategies reviewed in the MA: "property right changes" and "recognition of gender issues" (MA 2005b). Although few people would dispute that assigning unambiguous ownership of resources and assuring equality of opportunity are socially desirable goals, neither response was rated fully "effective" by the Millennium Ecosystem Assessment. One reason cited was that neither response in and of itself assures the enhancement of the targeted ecosystem services or improvements in human well-being. If these strategies are to be effective, it is necessary to ensure that those assigned property rights in natural ecosystems attach greater value to preserving ecosystems than to converting them to other uses (see Box 6.1). Similarly, if recognition of women's rights and interests is to be an effective response to ecological degradation, it must be ensured that the women empowered as a result will choose preservation over other uses (see Box 6.2). If these assumptions are not met, then these strategies—desirable though they are for a number of other societal reasons—cannot be assessed as effective for ecosystem service preservation.

In deciding between the institutional and behavioral approaches of the second tier and the instrumental strategies of the third tier, potential actors will want to devote special attention to the alignment of awareness with power. It may well be that institutional or social reform is essential for the achievement of ecological goals, but some actors devoted to the achievement of ecological goals will lack the capacity and access to implement such reforms.

Figure 6.3 summarizes a decision process for determining appropriate intervention strategies. It involves answering three questions: "Is the problem well understood?" "Are institutions and governance adequate?" "Are social conditions

Box 6.1. Farm forestry in India

In the mid-1970s, policymakers across the world perceived an impending fuelwood crisis caused by rapid increases in population (especially in rural areas in poor countries) accompanied by a decline in area under forest cover. The Indian government responded to this crisis by introducing social forestry programs on a large scale. One element of these programs was to provide incentives for small and marginal farmers to cultivate fuelwood trees on their own lands, for their own use, thereby relieving pressure on remaining natural forests. This "farm forestry" program was promoted through the provision of subsidized seedlings and technical support by government forest departments.

Subsequent evaluations suggested that the farm forestry program was adopted by farmers, but typically by large farmers who perceived trees as a valuable cash crop (as raw material for the paper and matchwood industries), not by small and marginal farmers (for whom growing trees was not the best option on their limited landholdings). The desired outcome, which was to be a decline in pressure on natural forests for rural fuelwood needs, was not achieved, although there was a substantial increase in tree cover (but typically of species that had a market value). The intervention was unsuccessful in alleviating pressure on natural forests as a source of rural energy supply, so it failed to deliver benefits for either the natural ecosystem or the well-being of poor rural households.

Box 6.2. Water, work, and women in rural India

In the Indian state of Gujarat, women spend long hours of backbreaking labor gathering and carrying water to their homes. The quality and reliability of water supply could be made considerably better by instituting ecological and engineering interventions. Traditionally, however, because women did not have social standing or political power, these improvements often were not realized.

In 1986 the State Water Board invited the Self-Employed Women's Association (SEWA) in Gujarat to help improve the water supply in rural areas. SEWA initiated the Water, Work and Women program. Local groups known as *pani samitis* ("water users' committees") cooperate to maintain and manage local water sources. The program has both required and motivated social change. To give one example, some local men opposed their wives' or daughters' involvement in the program because it required women from different castes to interact. One of the many consequences of the institutional and social reform that relaxed such restrictions has been the improvement of water supply.

Although SEWA and its Water, Work and Women program have achieved great results both in providing water to communities and advancing the rights of women, practitioners concerned with the ecological consequences of such institutional reforms should think broadly about such matters. Watershed protection and restoration—the preservation of ecological services—is just one of several activities conducted by the *pani samitis*. They also line ponds with plastic to prevent salinization, install pumping and piping, and make other improvements. Moreover, among the consequences reported is an increase from one to three crops planted per year in areas with improved water supply. While such developments testify to the tremendous social and economic benefits of the program, practitioners focused specifically on improving natural ecosystems and their resilience may want to consider the broader effects of such institutional strategies.

Source: Kapoor 2003, Panda 2007.

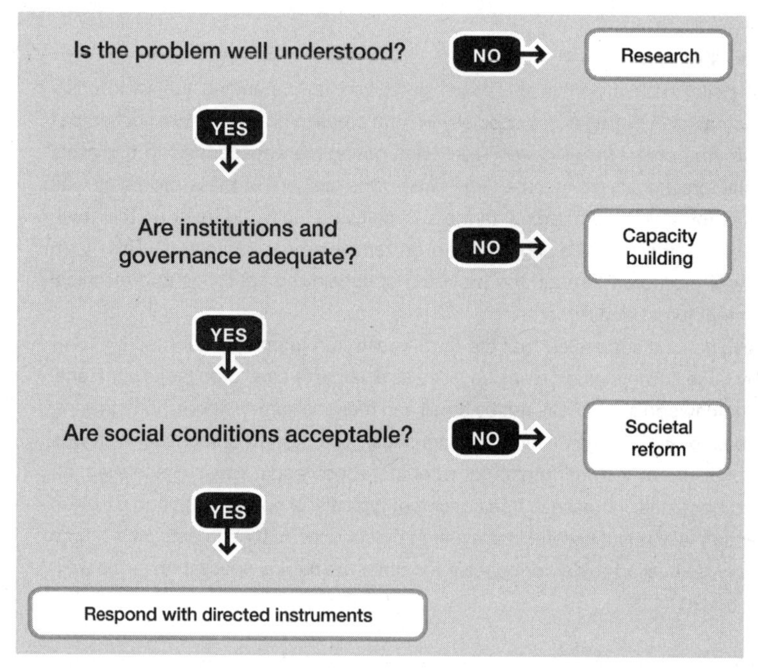

Figure 6.3. A decision process to choose types of intervention strategies.

(especially participation and equity) acceptable?" If these questions can be answered affirmatively, practitioners may go on to crafting specific instrumental strategies. If they cannot be answered in the affirmative, practitioners must decide if they have the competence and power to intervene at a more fundamental level, or if they should instead focus their efforts elsewhere.

The implicit starting point of the analysis is the perception that there is a problem: ecosystem services are being lost, with consequent impacts on human well-being. Then the first logical question is, "Is enough known about the problem to devise a response to it?" There have been innumerable instances in which the answer to that question was "No" (see Box 6.3).

But if problems are well enough understood so as to know what strategies would be appropriate in theory, the next question to ask is, "Can effective strategies be adopted in practice?" Implementing strategies typically requires some social apparatus with which to distribute rewards or mete out punishments. This could be done in a formal and centralized fashion, as might be the case with payments made from an agricultural or environmental ministry for the provision of ecosystem services from forest or farmlands, or in a less formal, decentralized way, as in the various structures traditional communities have developed to assure compliance with community norms for land and resource use. In the absence of some such governing institutions, however, as might occur in frontier areas or in regions enmeshed in conflict, it may be difficult or impossible to motivate more ecologically benign practices.

Closely related to the notion of adequate institutions and governance is a concern with social conditions, especially issues relating to participation and equity. It is

Box 6.3. The importance of understanding the problem: The case of fire management

To many people the appropriate response to fire in ecosystems designated for protection has been to prevent its occurrence and to extinguish fire once it has been detected. In a number of places, however, the wisdom of such policies has come into question. Lucas et al. (2008) write:

> Since the 1970s environmental managers in Canaima National Park, Venezuela have attempted to eliminate local savanna burning practices due to a widely held belief that it causes a gradual reduction in park forest cover. Apart from attempting to teach the native Pemon new fire burning techniques, for decades resource managers have tried to introduce new farming alternatives in order to reduce the reliance in traditional burning practices, but managers have had little success. A recent study, however, has shown that far from being a destructive activity, Pemon savanna burning is in fact an important tool to control large destructive fires. The failure of resource managers to understand the purpose of Pemon fire burning practices and value local environmental knowledge has been a key source of conflict. These failures are responsible for contributing to important alterations in local fire burning regimes which increase rather than reduce the occurrence of large scale destructive fires in the park.

Similar histories of fire suppression followed by a renewed appreciation for traditional fire management policies have been observed in many other areas.

possible that institutions exist through which local people can be compelled against their will to protect more of the ecosystems around them than they would otherwise choose to. If the focus is on human well-being and its relationship to ecosystem services, however, steps to preserve the latter should not unjustly undercut the former. As the adverb "unjustly" suggests, determining when this might occur often involves the exercise of subjective judgment. Preserving ecosystems that provide benefits to a community generally requires that some members of the community forgo uses that would be more privately rewarding but less beneficial to the community as a whole. It seems reasonable to suppose that people who have been traditionally engaged in uses that are now to be restricted, or who have few if any alternative means of livelihood, should be compensated. Such choices often require that "legitimate" claims for compensation be distinguished from others, a process that may require drawing somewhat arbitrary lines between claimants (Brandon 2002).

If practitioners understand the problem being faced and can work with capable institutions to implement strategies in an equitable fashion, they should begin to apply the criteria we will discuss in Section 6.2 to choose among the instrumental strategies described in Section 6.3. But if they cannot answer the questions posed in Figure 6.3 affirmatively, they may face difficult choices. For people in some positions it is natural and appropriate to conduct essential research, devise and improve institutions of governance, and work to enhance participation and equity. For those whose expertise and capabilities lie in these areas, the enhancement of ecosystem service benefits may only be a fortuitous by-product of their pursuit of the primary goal of research or governance. For those whose fundamental interest lies in preserving and enhancing ecosystem services, however, the identification of institutional impediments to achieving their goal may constitute a major roadblock.

The Bohensky and Lynam paradigm of impact scope, awareness, and power may be useful in charting a course. Practitioners may have become aware that problems within a certain scope of impact are detracting from human well-being, but they must then ensure that they, or a viable partner, have the power required to bring about change. Each of the "No" branches in Figure 6.3 directs the reader to a response to overcome the problem: "Do research." "Build capacity." "Reform society." These are not easy tasks. This chapter cannot address the myriad possibilities and challenges involved in these deliberations, as the focus is more on instrumental strategies and the consideration of trade-offs in their adoption. Practitioners should carefully consider their own objectives, capabilities, and demands before seeking to address ecosystem service loss through institutional interventions. A fundamental question arises in assessing the efficacy of any response: "Is this where we should be putting our time, effort, and money, or would those resources be more effective elsewhere?"

6.1.3 How to identify binding constraints

In developing guidance on intellectual, institutional, and societal prerequisites for various intervention strategies, it is useful to think about constraints and control variables. The latter are quantities that can be changed continuously over some range in pursuit of better outcomes. Governments could, for example, vary the payment made to local landowners for maintaining landscapes that provide ecosystem services. Constraints are limitations on the values the control variables can take. Landowners cannot receive less compensation for providing ecosystem services than they would receive by converting the land to alternative use, for example, as they would then pursue the alternative.

Although constraints are rarely so clear-cut as this—even in the example given, landowners could be compelled to preserve ecosystems even if they might otherwise choose not to—practitioners face a pragmatic choice of what instruments to treat as control variables and which considerations to treat as binding constraints. In other words, they must decide what factors they should treat as constraints on their range of action and which they may vary in trying to achieve their ends.

It is often possible to anticipate potential hurdles and to assess whether these are binding constraints that could potentially jeopardize the response or instead could be avoided by judicious interventions that ensure the success of the chosen approach. Table 6.1, based on the MA, suggests that such an analysis will need to assess binding constraints across four domains: political, institutional, economic, and social.

Table 6.1 argues that decision makers need to be realistic about the extent to which their interventions can fit within the socioeconomic, political, and cultural parameters of the society in which they are being proposed. And they should not attempt to implement strategies that are not congruent with the inflexible circumstances of the intervention context.

On the other hand, where conditions are conducive, changing some of these circumstances might be necessary as a precondition for the implementation of a chosen response. In this sense, the constraint is not binding, but identification of the

Table 6.1. Binding constraints—considerations that might rule out a potential response

Domain	Issue	Evaluation
Political	Can all potential stakeholders be identified?	Not likely to be a binding constraint, unless neglected stakeholders mobilize political opposition
	Is the political context supportive?	If not, could be a binding constraint
	Can the political context be changed?	If not, a binding constraint
Institutional	Is there adequate capacity for governance at an appropriate scale?	If not, could be a binding constraint
	If not, can adequate capacity for governance be created?	If not, a binding constraint
Economic	Is the outcome cost effective?	Could be a binding constraint if funds are limited
	Are there secure and well-defined property rights?	Could be a binding constraint if there are numerous competing demands on the resource
Social	Is the outcome equitable, in a distributional/material sense?	Could be a binding constraint, if this is a high priority or if those disadvantaged by the response can effectively oppose it
	Does the outcome violate the cultural norms of particular groups?	Not likely to be a binding constraint unless consensus is an explicit objective

Source: MA 2005a.

constraint allows the decision maker to create enabling conditions for the chosen response to work more effectively. Box 6.4 presents an example in which an economic constraint that might otherwise have been binding was eased by combining economic and social objectives.

6.2 How to resolve trade-offs

Section's take-home messages

- Most intervention strategies involve winners and losers. Practitioners have to make judgments that balance competing interests when evaluating intervention options.
- Multi-criteria analysis allows different criteria to be considered in making a judgment about the desirability of an action, and it is also useful when trying to incorporate the views of diverse actors on an intervention strategy.

Box 6.4. Working for water

As is the case in many areas of developing countries, parts of South Africa suffer from a shortage of clean water. Such shortages are in many instances exacerbated by the presence of alien species. Nonnative trees sink deep roots into the soil, extracting limited supplies of groundwater.

The South African Subglobal Assessment reports on efforts to remove exotic species and restore native ecosystems and their associated hydrology under South Africa's "Working for Water" program. The process of removing nonnative species is extremely labor intensive and consequently very expensive. This expense might have been a binding constraint preventing implementation of the Working for Water program and thus the restoration of native species and the increased availability of water. But South Africa also faces serious social problems arising from inadequate employment opportunities, particularly for traditionally disadvantaged groups. By hiring such people to clear land of exotic vegetation, South Africa was able to restore ecosystem services and enhance job opportunities.

While "win-win" outcomes are certainly desirable, generalizing from such examples should be done only with care. Two features of the Working for Water program may distinguish it from some other public works efforts. First, the benefits arising from clearing land of nonnative species are largely public goods: they accrue to people other than those who own the land being cleared. Second, there were severe problems of unemployment and underemployment in the regions where the Working for Water program was initiated. It may have made good economic and social sense in South Africa to pay more to employ workers to clear land than the market value of the land once it was cleared. Before emulating such a program, however, practitioners in other areas should ensure that the activities they contemplate serve similar public purposes.

Source: SAfMA 2004.

- Cost-benefit analysis reduces different outcomes to a common unit of measurement, but it is controversial because of several problematic methodological assumptions. Despite this, it remains a pragmatic and popular technique that provides helpful information for decision making.
- If a full cost-benefit analysis is not possible, cost-effectiveness analysis can be used to assess whether a planned intervention strategy can be achieved at a reasonable and acceptable cost.

The assessment of most actions that might be contemplated to address the degradation of ecosystem services will entail an analysis of trade-offs. Almost any public policy choice creates winners and losers (see Box 6.5). To improve one measure of well-being, there typically must be some deterioration in another measure. To make one person better off, typically some action is taken that will make another person worse off. How should society make such choices? A variety of solutions have been proposed.

Various criteria might be considered in making choices about the environment: the welfare of the people who favor a change from the status quo, the welfare of those who oppose it, and in some formulations the welfare of other organisms or

Box 6.5. Visualizing trade-offs

The subglobal assessments conducted by the Alternatives to Slash-and-Burn (ASB) program have adopted a useful way of visualizing trade-offs. They assemble matrices whose cells reflect the "payoffs," in physical or monetary terms, to different stakeholders from different policies. The following table provides an example from the forest margins of Sumatra.

	Beneficiary or interest					
	Global community		Agronomic sustainability	Nation		Local
Land Use	Carbon sequestered (tons/ha)	Plant species/ standard plot	Production sustainability rating (0–1)	Returns to land ($/ha)	Labor input (days/ ha/yr)	Returns to labor ($/day)
Natural forest	306	120	1	0	0	0
Community-based forest management	136	100	1	11	0.2	4.77
Commercial logging	93	90	0.5	1,080	31	0.78
Rubber agroforest	89	90	0.5	506	111	2.86
Oil palm monoculture	54	25	0	1,653	108	4.74
Upland rice/ bush fallow rotation	7	45	0.5	–117	25	1.23
Continuous cassava degrading to *Imperata*	2	15	0	28	98	1.78

Source: ASB 2003.

Not surprisingly, the ecosystem services that provide the broadest global benefits—carbon sequestration and species diversity—yield the lowest payoff to the national governments and local people whose actions determine the state of affected ecosystems.

Such a representation of trade-offs can serve a couple of purposes. The first is simply that it makes clear, and in some instances stark, that there are, in fact, trade-offs to be made. Second, such representations may be the basis of "what if" exercises. For example, if Sumatran forest authorities could earn €25 per ton of carbon sequestered on the land they control, their incentives and choices would likely be very different.

It is important to be careful in doing such exercises, however. The ASB project suggests that doubling the return of nontimber forest products might actually reduce the area of forest maintained. Why do they get this counterintuitive result? Because local people might invest the proceeds of such sales in clearing more land.

Source: ASB 2003.

entities affected by the proposed change. Once practitioners accept that there is in fact a trade-off to be resolved among these different criteria, they have implicitly acknowledged that an analysis weighing multiple criteria is required. The question, then, is how to do it.

The simplest approach is to establish a hierarchy among criteria. One way to resolve confrontations between competing goals is to decide that one takes precedence over another. This leads to problems, however, if the starting position is that certain considerations are always more important than others. What can be done when facing competing imperatives? In practice, what is typically done is not to privilege any particular interest with absolute priority but rather to assign weights to different considerations. This section considers three alternatives that seek to balance competing interests: multi-criteria analysis (MCA), cost-benefit analysis (CBA), and cost-effectiveness analysis (CEA).

6.2.1 Multi-criteria analysis

Multi-criteria analysis refers to a procedure for aggregating different criteria into a single index to be used in making a complex decision. The technique attempts to integrate different elements of a decision, recognizing that these might not always be easily comparable. One way of doing this is through a process that either ranks or rates the different outcomes according to the desired criteria and then compares them using a formula. This formula could take any of several forms. For instance, it could be a simple checklist that would list several desired features. The "formula" in this case is "assign one point for each criterion met, and add up the points." Such a procedure is unlikely to prove robust, however. An option may fail to meet one criterion, while exceeding others in more-important measures.

MCA is a useful technique for aggregating the views of different actors on a particular issue. It lends itself particularly well to team-based decision making, where the preferences of different team members can all be integrated into the final decision without necessarily privileging the opinion of one person. This avoids the risk of more "dictatorial" techniques, in which those conducting the assessment impose their own sense of what is important in determining the well-being of the people and ecosystems being evaluated. In this sense, MCA encourages more participatory decision making.

In sophisticated uses of multicriteria analysis, it is common to define aggregate measures as weighted sums of individual numerical measures (see the discussion of the Human Wellbeing Index in Chapter 4). However, multicriteria analyses in which different attributes are weighted and summed to form a total are, at best, approximations of more accurate measures of performance that would be based on more complicated, nonlinear multivariate functions. This is not to say that simpler MCA procedures are ill advised or wrong; rather, only that any such procedure is only as reliable as the assumptions that underlie it.

6.2.2 Cost-benefit analysis

Cost-benefit analysis is a specific form of trade-off analysis that reduces the different outcomes to a common unit of measurement. Typically, certain quantities ("benefits") are multiplied by positive weights ("prices"), while certain other quantities ("costs") are multiplied by negative weights (also "prices"), and then all of these weighted quantities are summed to derive measures of the net benefits of the project or program under consideration. If net benefits are less than zero, they are "net costs."

CBA differs from many other forms of trade-off analysis in that the weights placed on quantities in the analysis are inferred from observed data rather than being assigned based on moral principles or *a priori* judgments. However, this "objective" attribute may be both a virtue and a defect of the procedure. It implicitly assumes that additional money income (or loss) is of equal moral worth regardless of the initial wealth of the recipient. This problematic assumption may be relaxed, but only at the cost of introducing other complications.

In the actual use of CBA, several issues arise; all are controversial:

- *What can be done to infer the prices of commodities that are not traded in markets?* This is the issue of nonmarket valuation and has been the subject of many articles and texts (see, e.g., Freeman 2002; see also Box 6.6).
- *How should nonmarginal changes be measured?* In economic theory, prices are related to marginal well-being: the improvement in welfare realized in response to a small change in the provision of a good or service. Cost-benefit analyses can be done in consideration of large changes (in fact, it only makes sense to do a CBA in consideration of nontrivial changes), but it typically involves the use of advanced statistical and/or computational techniques.
- *How can measurements be aggregated across people?* The economic theory underlying CBA describes how one person's welfare may change when circumstances change, but there is no good way to compare one person's welfare with another's. Conventional cost-benefit analysts typically compare monetary estimates of what different people would pay to adopt a certain policy. However, this procedure relies on the implicit assumption that money in the hands of one person is of equal moral worth to an extra unit of currency in the hands of any other. This assumption is deeply problematic.

Another very controversial aspect of cost-benefit analysis concerns discounting. This is the practice of assigning a lower value to costs or benefits that would accrue at some point in the future than would be assigned to the same costs or benefits that would accrue immediately. The argument for adopting some form of discounting relies on three elements:

- People have an intrinsic preference for "sooner" rather than "later."
- Capital is productive. Benefits forgone today in order to increase production next year will lead to a more than one-for-one increase in future production over current consumption sacrificed.
- People grow wealthier over time. If current investment pays off in increased future consumption, that consumption will be less valuable to its wealthier (future) recipients than the same quantity of consumption today.

For these reasons, it is generally considered appropriate to discount costs and benefits at a constant exponential rate when evaluating projects and programs.

Box 6.6. Measurement of nonmarket values

Cost-benefit analyses involving environmental improvements or ecosystem services are very difficult because damages to the environment and ecosystem services typically do not have market prices that can be used to measure the value to society. This is not necessarily because these things are not valuable, but rather because they are public goods that accrue to many people. Economists have developed a number of techniques of inferring values for goods that are not traded in markets.

Revealed preference methods rely on observations of behavior in related markets. They include:

- Hedonic pricing, which involves making inferences concerning the value of component attributes of a market good from the price of the good. For example, the price of a house reflects ecosystem services from which it benefits. It often requires considerable subtlety to disentangle the effects of different attributes, but sophisticated statistical techniques have been developed to address these issues.
- Travel cost methods, in which peoples' desire to travel to recreational or scenic sites is assumed to be related to ecological attributes of the sites. As admission prices are often not charged, the willingness to pay for the attributes of such sites may be inferred from the "price" people pay, in time and travel expense, to visit them.
- Production function approaches, which involve treating ecosystem services as unpriced inputs into the production of agricultural or manufactured products. The "price" of the ecological input can then be imputed from its contribution to the production of marketed commodities. Again, subtle statistical treatments are often required to disentangle the effects of inputs.

Stated preference methods have been developed to estimate the value of things that are enjoyed even when they are not consumed or used. People may care about the preservation of wild species such as mountain gorillas or giant pandas, for example, even if they never expect to buy products made from these animals, visit their habitat, or even see them in a zoo. Under such circumstances, since there is no market transaction to which people's value of their existence may be related, the only way to estimate the worth people place on them is to ask questions on surveys. Such "contingent valuation" or "conjoint analysis" exercises are extremely controversial. Their expense and credibility should be considered carefully before they are undertaken.

Conventional wisdom has it that appropriate discount rates range from between approximately 3% and 10% per annum, with the lower rates more appropriately applied to wealthier, more stable economies and rates toward the higher end of the scale applied in the developing world.

With its philosophical foundations in considerable doubt, CBA is justified largely on the pragmatic grounds that it is something that analysts can do, even if it remains less clear why its results should be considered (Posner 2006). It is often best treated as one (among many) inputs into decision making rather than the sole basis for making policy choices. CBA can be a powerful and useful tool, and it certainly has much to recommend it to the analysis of past and prospective decisions. But there are some subtle points to be appreciated, and practitioners wishing to employ cost-benefit analysis in the assessment of responses should bear certain caveats in mind (see Box 6.7).

6.2.3 Cost-effectiveness analysis

The objective of cost-benefit analysis is to total all the costs and all the benefits of a proposed action. If total benefits exceed total costs, the action should be carried out. In many instances, however, it is far easier to estimate the costs of an action than to quantify its benefits. Consider, for example, choices made to preserve endangered species. The cost of land purchased as a wildlife reserve or income forgone from prohibiting incompatible activities is often easily calculated. But what is the monetary value of the benefits of species preservation?

When the answers to such questions are especially difficult, economists often settle for the less demanding alternative of performing a cost-effectiveness analysis. This asks if a specific objective (preserving an endangered species, for example) is accomplished at the lowest possible cost. The CEA may be sufficient to establish that the costs of securing certain unquantifiable benefits are "reasonable." And, of course, conducting a CEA can help determine the least expensive way of achieving

Box 6.7. Common errors in economic analyses

While many of the ways in which programs and projects may be assessed are straightforward and make common sense, common errors should be avoided:

- *Marginal versus total values.* Economic value is determined by how much an additional amount of a thing is worth, not how much the thing is worth in total. If an ecosystem service is to be reduced but not eliminated, the loss to be estimated is the benefits forgone as a consequence of the reduction.
- *Value-added.* The argument is often made that developing countries should capitalize on their natural resources by engaging in "value-added processing" at home rather than simply being suppliers of raw materials. This confuses the sources of value. Value arises from scarcity; if the resources a nation controls are not scarce, and hence economically valuable, the country cannot generate real profits by investing in processing facilities unless it is particularly favored in processing. In many instances "value added" is simply recouping costs incurred to engage in processing.
- *Substitutes.* If there are alternative ways to generate the goods or services of natural ecosystems, the value of such goods and services cannot be greater than the cost of the alternative.
- *Replacement costs.* It follows from the above observation that if there are cheaper ways of producing a good or service than replacing the system that currently provides it, the cost of replacement will overstate the value of the good or service.
- *Double-counting.* There are often many ways of estimating economic values (see Box 6.6). Calculating values by different methods is sometimes useful to check on against the other, but it is important not to count the same value twice.
- *Alternative metrics.* While embodied energy, ecological footprints, and other physical measures may be useful for some purposes, they generally cannot be used in economic valuation.

social goals (such as when decision makers are faced by a choice of alternative strategies).

6.3 How to confirm that the right strategy was chosen

Section's take-home messages

- An intervention strategy can be judged to be effective if it achieves its ecological objectives without unduly compromising other ecological or social objectives.
- A number of instrumental responses have been used, and each is associated with advantages as well as potential pitfalls.
- Community management works well for some local resources and may be either formal or informal; however, it is not necessarily an equitable or egalitarian system, and it may not always be easy to introduce through external intervention.
- Command-and-control regulation may be appropriate at larger scales, but it relies on effective enforcement and implementation.
- Market-based incentives include both tax- or fee-based systems and quota or cap-and-trade systems. These need to be carefully planned and implemented in order to ensure effectiveness.
- Payment for ecosystem services programs are innovative theoretical ways to reward providers of valuable ecological services, but there are few working examples that truly involve direct and contingent payment systems in practice.
- Sustainable use strategies are innovative ways to ensure that conservation remains compatible with economic uses, but they require well-designed and credible certification programs to be effective.

To evaluate particular response options, we must address three questions: First, by what standard or standards are actions to be judged? Second, what types of responses ought to be considered? Third, how effective have such responses been or might they be? This section discusses these questions in order.

6.3.1 How to define "effectiveness"

We adopt the Millennium Ecosystem Assessment's definition of "effectiveness." Defining "responses" as "human actions, including policies, strategies, and interventions, to address specific issues, needs, opportunities, or problems," the MA defines effectiveness as follows:

> A response is considered to be *effective* when its assessment indicates that it has enhanced the particular ecosystem services (or, in the case of biodiversity, its conservation and sustainable use) and contributed to human well-being without significant harm to other ecosystem services or harmful impacts to other groups of people. A response is considered *promising* either if it does not have a long track record to assess but appears likely to succeed or if there are known means of modifying the response so that it can become effective. A response is considered *problematic* if its historical use indicates either that it has not met the goals related to service enhancement (or conservation and

> sustainable use of biodiversity) or that it has caused significant harm to other ecosystem services [MA 2005b:123; emphases added].

We adopt the MA's view that effectiveness is generally to be measured against the goal of "enhanc[ing] . . . particular ecosystem services." We also note another element of the MA definition: an approach is effective if it achieves its ecological objective without unduly compromising other ecological or social objectives. This latter criterion raises the issue of trade-offs between the achievement of desirable objectives, a topic already discussed in some detail.

6.3.2 How to identify response options

To identify the most effective responses to ecological degradation, it is important to understand the range of possible options. This chapter focuses on "instrumental responses": actions taken specifically and explicitly to address deterioration in or threats to ecosystem services. As noted earlier, a variety of other responses might be undertaken to create the preconditions for implementation of instrumental responses. The efficacy of such enabling responses might be measured by the extent to which they permit the successful implementation of instrumental responses.

This section briefly considers five broad categories of programs to address the degradation of ecosystems services and biodiversity: community management, command-and-control regulation, market-based incentives (MBIs), payments for ecosystem services (PES), and sustainable use.

Community management

In some cases, public goods such as ecosystem services can be better managed if the "public" who benefit from their provision manage the services cooperatively. Communal ownership often makes sense when critical components of an ecosystem are not easily confined to a particular area, as in the case of many animals, or when it is important to maintain large contiguous areas in order for services are to be maintained, as is often the case with forests, grasslands, or wetlands.

Successful common property or community management regimes tend to be localized; it may either be formal and recognized by the state or operate as informal, traditional systems based on the (benign) neglect of the state. Access is allowed for all members of the community, and nonmembers are excluded. Under such circumstances, an effective response may be for communities to adopt rules and norms for managing the ecosystems that provide them with services. Pioneering work by Elinor Ostrom (1990; see also Baland and Platteau 1996) provides guidance as to the determinants of effective common property management responses.

These conditions arise largely from a consideration of repeated interactions between community members. One person's maintenance of community infrastructure, or forbearance from the depletion of community resources, is more likely to be rewarded by another's reciprocation if both parties anticipate interacting over many years. Such understandings are likely to be upset if new residents arrive in traditional communities, and this has been a common problem. It is also increasingly acknowledged that communities are not simple homogeneous groups that work harmoniously to promote group objectives. What has emerged in the contemporary literature is a more subtle and nuanced understanding of communities as complex, dynamic, and characterized by internal differences and processes (Leach et al. 1999, Agrawal and Gibson 1999). Box 6.8 provides an example of how the decline of common property regimes can result in the degradation of ecological resources, while the restoration of communal rights may revitalize affected ecosystems.

Among the policy options open to conservation practitioners and their funders is to "encourage more effective common property management." It is important to ask whether outside actors can induce local people to do something that they have not yet found in their interest to do for themselves. Responses intended to promote more effective community property management can only be expected to improve biological conditions if communities are unaware of the effects of their actions or if individual decision makers do not take account of the reciprocal effects they have on each other. A community that has already solved its resource management problems to its own satisfaction will not be affected by appeals to work together if they are already working together. If a concern with *global* ecosystem services such as carbon sequestration or biodiversity protection motivates the response, the community can be expected to react to such a concern only if their incentives to do so are augmented by outside rewards.

Box 6.8. Van Panchayats in Kumaon, India

Communities in the Central Himalayas of northern India had been managing local forests for centuries when the British took control of the country. The forest ecosystems managed by villagers provided a variety of ecosystem services: water flow regulation and erosion protection, fodder for animals, fuel for cooking and heating, and construction materials. Elaborate systems had evolved over time for the maintenance and management of such forests. The rights of local villagers to forest products were closely regulated, and outsiders were prevented from unauthorized access to forests.

During the nineteenth and twentieth centuries, demand for wood increased, and new forest regulations were adopted to promote planting and harvesting pine rather than oak species. While the native oak provided the suite of ecosystem services just mentioned, pine was useful for timber and resins, but not for fodder or erosion control. Moreover, local villagers were generally precluded from benefiting from the sale of pine timber. Consequently, local people exercised less care in the management of forests, and the forests deteriorated. Moreover, new regulations made the mistake of assuring "all bona fide inhabitants" of a region access to forests for traditional uses such as grazing. It was then impossible for inhabitants of one village to exclude those of another from exploiting common resources or to punish those found to be overusing resources or failing to contribute to their maintenance.

Forest degradation was partially reversed with the adoption of *van panchayats* (local management committees) in the early twentieth century. Under these institutions, local people were again given rights to the proceeds of their forests and a greater hand in their management. Equally importantly, they were again allowed to exclude unauthorized users from exploiting their resources. Some of these institutions remain robust today, almost 80 years after their introduction.

Source: Somanathan 1991.

Command-and-control regulation

When public concern (by which we mean to encompass local, regional, national, and global concerns) with the management of ecological resources reaches a critical level, public action is taken. The enactment of statutes and regulations as a response to ecological degradation is not necessarily a distinct approach from community management but rather an extension of them using the mechanisms available to larger polities. Larger communities have less of the direct person-to-person interaction on which traditional community management regimes depend for monitoring individuals' performance and dealing with those who violate community norms. Larger communities do, however, have more organized regulatory apparatus and police power to ensure compliance.

Hence, large modern communities have developed a variety of regulatory means by which to restrict ecological degradation. The traditional ways to do this have relied on the police power of the state to enforce command-and-control regulation. The state commands that individuals and firms observe certain controls. There need be nothing authoritarian or undemocratic about such measures. The majority may recognize that it is in their common interest to prevent the degradation of their environment, and they may introduce rules that enforce appropriate restrictions on individual actions.

The form of state control most germane to the protection of ecosystem services and biodiversity is the protected area. Establishing protected areas can be an effective response to the degradation of natural ecosystems. National parks and other protected areas are, by and large, respected and maintained in the world's wealthier nations. In poorer countries, however, protected areas are sometimes derided as "paper parks," set aside in theory only. The success of prohibitions on the exploitation of protected areas will depend on a variety of factors. The most obvious is the financial and political resources expended on preventing intrusion. Perhaps equally important, however, is the legitimacy with which protected areas are regarded by local populations. For both moral and pragmatic reasons, local people should be compensated for their loss of access to protected areas.

Market-based incentives

There may be more effective ways to control ecological degradation than by prohibiting certain actions. It is not always necessary to require everyone to stop taking actions that degrade ecosystems. If some can stop at a lower cost than can others, it may make sense for them to bear a greater share of the cost of forgoing degrading uses. Of course, the lower-cost entity will only assume a greater share of the cost if it is compensated for doing so.

This is where market-based incentives are useful. Given that a certain amount of ecological degradation must be borne in exchange for social benefits, there are several ways to achieve the efficient allocation of degradation across actors. They involve setting a "price" for degradation and allowing actors to choose how much degradation to "buy" at that price. The price of degradation could be set with a tax or a fee. Or a price will emerge if regulators place a limit on the total amount of degradation permitted and allow actors to trade among themselves the right to engage in it (a cap-and-trade system).

There are two chief differences between the tax or fee-based systems and the quota or cap-and-trade approach. The first concerns the distribution of the burden. With taxes and fees, those who degrade the environment pay the state. With tradable permits, one polluter purchases from another the right to emit. The state can, however, collect revenues in a tradable permit program by auctioning permits to pollute rather than by giving them away for free.

The second key difference between taxes and tradable permits is that the effects of the latter are more certain. In such a program, authorities set a fixed quantity of allowable degradation, and they allow private parties to trade among themselves to determine how much degradation each causes. With taxes, there is no predetermined limit on how much degradation can occur; this will be determined by the economic circumstances of those who benefit from the degradation.

The most common application of MBIs has been to industrial pollution; the two most widely cited tradable permit programs are the U.S. sulfur dioxide trading program and the European Union's greenhouse gas Emissions Trading Scheme. Industrial emissions have important consequences for ecosystem services—as demonstrated perhaps most prominently in the case of greenhouse gases—but often for ecosystem services the source of concern is the depletion of particular biological resources or the conversion of habitats providing the services. There have been experiments with market-based incentives in these contexts as well. Various nations and jurisdictions have experimented with tradable development rights—the right to convert land in one area in exchange for an obligation to protect it in another—and the Business Biodiversity Offset Program (BBOP 2008) is an attempt to establish a market in biodiversity protection. In Brazil, legislation has been in place for many years requiring that a certain fraction of original forest be maintained in many areas; in recent years, steps have been taken to initiate trades in forest preservation (see Box 6.9).

But a variety of factors must be considered in evaluating the efficacy of such MBIs in general, and these considerations are often particularly important in the context of ecosystem services:

- *Market-based incentives are only as reliable as they are credible.* Commitments to preserve ecosystems in one area in exchange for valuable considerations elsewhere must be credible. They must be monitored and enforced. Nations, regions, or individuals who cannot provide evidence of compliance will either be excluded from participation or will undercut the system.
- *Market-based incentives must deal in a "common currency."* Desirable economic efficiency effects of MBIs are more likely to be realized if there is a larger market over which compliance and, in the case of tradable permits, trading can occur. On the other hand, however, wider possibilities for trade increase the likelihood that such trade will exchange "apples for oranges." Trading ratios for disparate assets must be established if MBIs are to be used over broad areas.
- *Market-based incentives must guard against "leakage."* One danger of MBIs is that they will just displace degrading activities to other areas. Suppose that a system of tradable permits is established by which landowners in a region are prevented from deforesting more than a certain fraction of the land they own. This will reduce the supply of forestland relative to the demand for timber and

Box 6.9. Tradable forest rights in Brazil

In many parts of Brazil, landowners are required to maintain 20% or more of their holdings in native forest. The expense of doing so may vary dramatically between different areas. At the same time, however, the ecological consequences of forest preservation may vary greatly between the areas of forest preserved or restored.

Chomitz et al. considered these issues in a study of the potential for tradable forest rights in Brazil. There is a trade-off in expanding the area over which trading in forest rights is allowed to occur. On one hand, the larger the area, the greater are the expected cost savings: it is more likely that someone who can preserve forests at a very low opportunity cost will accept the obligation to do so. On the other hand, the larger the area over which trades are allowed, the more likely it is that disproportionate areas of forest will be traded.

In the state of Minas Gerais, the costs of complying with forest-maintenance regulations could be reduced substantially if trades were allowed to occur over a wider region. Surprisingly, the study's authors also conclude that the ecological benefits of a wider trading program could be greater than one in which conservation or restoration must occur on-site. The reason is that forest restoration in already degraded land would likely not be very effective. Moreover, if some of the cost savings under the trading program could be directed to particularly sensitive areas, the results might be still better.

The authors point out that Brazil is unusual, however, in that there already exists a quota set by law on deforestation. They suggest that it might be more difficult to initiate a tradable forest rights program entirely from scratch.

Source: Chomitz et al. 2004.

farmland. The price of land and timber will go up. Unless the same restrictions can be imposed on all forestland in the area, other lands may be exploited, weakening the impact of the conservation program. MBI programs must be carefully planned to ensure that alternative, equally sensitive assets are not degraded when one area is designated for protection.

Payments for ecosystem services

The ultimate evolution of market-based incentives regarding the degradation of ecosystem services is a system under which payments are made for their preservation. Such systems have emerged in recent decades in the policies of both wealthier nations and some developing countries. While the theoretical case for payments for ecosystem services is impeccable, considerations often arise in practice that limit their application or at least underscore the preconditions that must be met for their adoption.

The most important consideration is that ecosystem services remain, by and large, public goods. Consequently, the problem of "free ridership" arises: any payment one person makes to protect ecosystem services benefits many people. For this reason, conventional wisdom has it that public goods must be provided by public action: the state must compel individuals to pay their fair share via taxes, fees, or similar measures. This is not to say that this process must be undemocratic or unjust; to the contrary, the nations in which PES systems have caught on tend to be progressive. The point is only that PES, like most other responses to ecological degradation, requires that public action be mobilized to address the problem. Moreover, inasmuch as different ecosystem services affect different elements of the global community, truly effective action to address global ecosystem services requires a global commitment of resources. This is not necessarily impossible—in fact, proposals have been made recently for International Payments for Ecosystem Services (OECD 2003, OECD 2004)—but it should be appreciated that such programs require an unusual level of international cooperation.

Ideally, PES programs would be initiated whenever the benefits provided by the preservation of ecosystem services outweighed the costs of preserving them. In practice, however, PES programs have often arisen when political circumstances align constituencies behind them. For example, the Conservation Reserve Program (CRP) conducted by the U. S. Department of Agriculture has set aside over 15 million hectares of farmland—almost 4% of the nation's total agricultural land—for habitat preservation. The CRP has proved successful in preventing erosion, reducing sedimentation, protecting wildlife, and providing other ecosystem services. Political acceptance of the program is surely enhanced by the fact that it directs monetary payments to a powerful constituency—U. S. farmers—that had already established claims on public support via other agricultural price and income support programs. One potential downside, then, of PES programs is that they may not necessarily devote resources to the areas of greatest ecological merit but rather to those with a combination of desirable ecological attributes and political interests.

A related concern is that PES programs face a trade-off between administrative complexity and ecological effectiveness similar to the one described for market-based incentives. A program that simply announces payments of a certain amount for land will enroll the least expensive land, which may or may not be valuable for its ecological attributes. Inasmuch as inexpensive land tends to be located far from concentrations of population and economic activity, however, it cannot be expected to provide locally valuable ecosystem services. On the other hand, careful evaluation of the ecological potential of parcels proposed for conservation may lead to better selections. But there are two pitfalls here. First, such careful evaluation is expensive in terms of administrative time and effort. Second, the more complex and arcane a system for evaluating conservation land becomes, the more vulnerable it may become to manipulation for private gain or political advantage.

While PES programs ought to be initiated when the benefits of their enactment would justify the expenses paid to property owners, in practice a somewhat different calculation is often made. For example, Costa Rica's PES program holds out the possibility of generating reimbursement for carbon sequestered in forested areas. However, operating funds for the program are generated by a tax on fuels. So although there is a sense in which the conservation investments made under the PES program offset carbon emissions from fossil fuel use, there is no direct quid pro quo under which the beneficiaries of conservation compensate its providers.

This is not at all to say that PES programs are not effective or that they should not be included among potential options for protecting biodiversity and ecosystems. The point, rather, is that true international markets that involve direct and contingent payments for ecosystem services are only just beginning to emerge.

Sustainable use

A sea change in conservation strategy occurred in the last quarter of the twentieth century. For most of history, protected areas have been either maintained by common consent of the communities in which they are located or established and enforced by higher levels of government. In 1981, with the publication of the influential *World Conservation Strategy: Living Resource Conservation for Sustainable Development* by the International Union for Conservation of Nature, the U.N. Environment Programme, and the World Wide Fund for Nature, the focus began to shift (IUCN et al. 1991). A new emphasis on use, rather than simply conservation, of resources developed. "Use" was to be nonconsumptive, or at least sustainable, but under the new policy "natural areas" would be seen as productive economic assets rather than as reserves deliberately excluded from economic use.

The adjective "sustainable" implies constraints on the other uses that could be made of natural areas in order to assure a livelihood for their human inhabitants. Extensive deforestation and intensive large-scale agriculture, to say nothing of heavy industry, are obviously incompatible with "sustainable" use. If it were felt necessary to promote sustainable use as a conservation policy, it must have been because the alternative was perceived to be the devotion of previously unspoiled habitats to activities that were not consistent with their preservation.

Of course, someone must also pay the costs of monitoring required to ensure that certified products are, in fact, produced sustainably. This observation begs a couple of questions similar to those considered earlier. First, if consumers are willing to pay the extra cost of sustainably produced products, who should be responsible for providing certification? Obviously, there would be a conflict of interest issue in having every producer certify its own product, and this would carry over to having an industry group certify the products of its constituent members. It is not clear that such conflicts of interest necessarily prevent the operation of industry-funded certification systems, however. While concern has often been expressed as to "Who will certify the certifiers?" the value created by certification is only as great as its credibility. Companies would have little incentive to incur costs for certification that consumers did not trust.

It should also be remembered environmental certification is different in some respects from other forms of product certification and may, as a consequence, prove less effective. Consumers who purchase an electrical appliance whose performance and safety are certified by Underwriters Laboratories are concerned that they do not waste their money on a product that does not work or will endanger their family. People who stay at a hotel certified by Green Seal are concerned that their lodging choices do not have excessive negative consequences for the environment. These incentives may be greatly dissipated by the fact that the impacts of individual decisions are diffused across so large a public. So while certification may be a response to ecological degradation, it is unlikely to be effective without other accompanying measures. Broader measures, such as direct regulation of forestry, for example, might in turn obviate the need for certification. In short, certification measures may be useful and desirable responses in some contexts, but policy makers should consider carefully the interaction of these efforts with other responses, as well as their relative effectiveness.

6.4 How to link assessment of responses to decision making

Section's take-home messages

- Assessment of intervention strategies is a critical part of the decision-making and policy-learning process.
- Attitudes toward organizational learning shape the overall value of an assessment process and its impact on improving intervention strategies.
- There may be unexpected benefits from undertaking assessment processes, such as positive impacts on organizational culture, improved relationships with key stakeholders, and better communication outcomes, in addition to actual learning about the effectiveness of intervention strategies.
- Considerable methodological innovation has taken place in assessment processes, and tools now include a sophisticated combination of both quantitative and qualitative methods.

Both in an evaluative context (looking back at previously implemented strategies) and in an anticipatory context (looking ahead at possible alternatives), the real value of assessing alternate responses emerges once this gets fed into an iterative and continuous process of decision making and policy learning. Figure 6.4 illustrates the different stages at which assessment feeds in to the policy learning cycle.

An ex ante assessment is usually undertaken in order to understand the potential impact of a response on a desired objective; it is important to use forecasting and scenario planning approaches to generate the required information to predict possible impacts. This is important in the case of Strategic Environmental Assessments (now a requirement within the European Economic Area), Environmental Impact Assessments, Social Impact Assessments, and Regulatory Impact Assessments. Once a policy has been implemented, an assessment looks back to evaluate the effect of the policy on desired objectives; this might take place at the project level or at a more macro-level, within the context of sectoral or national reviews.

In the policy learning cycle, there is an important distinction between monitoring, which can be seen as a continuous process of data collection to track changes in specified indicators, and evaluation or assessment, which is a more reflective process that seeks to understand the nature of change. In this sense, assessing responses needs to go beyond a simple measurement exercise if it is to have any significant impact on the overall effectiveness of policy processes. However, attitudes to organizational learning vary, depending on the institutional context within which assessment takes place; hence, it is not always necessary that new strategies evolve out of a careful and empirically informed understanding of past experience (or of systematic scenario planning). If an organizational culture does not reward reflexive decision making, interventions may not necessarily be based on a careful assessment of past experience.

Some of the unexpected benefits of assessment processes that have been reported include process outcomes. The introduction of participatory monitoring and evaluation in some projects, for instance, might create an improved organizational culture that might benefit program delivery and effectiveness, even if the "data" from such an exercise are not all that valuable. The process of assessment might help create

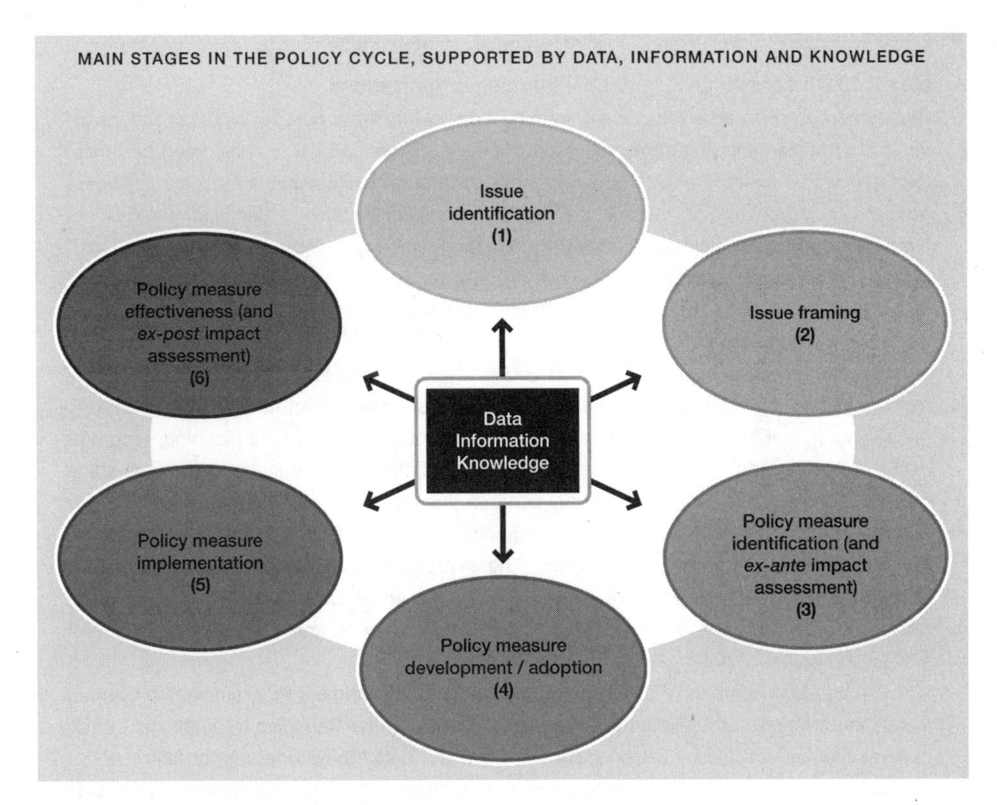

Figure 6.4. The policy learning cycle.
Source: EEA 2005

capacity and nurture a culture of shared learning on policy design, implementation, and evaluation (EEA 2005). Here, processes that allow affected stakeholders to engage with the assessment, as well as to shape its methods, might enhance the legitimacy of the process and create more inclusive learning outcomes, with lasting value beyond the specific context of the assessment.

In a learning organization or policy context, assessment is a critical part of the process, helping decision makers to assess the effect of planned interventions as well as to restructure and redesign ongoing programs in an iterative cycle. There is, however, often a considerable disjuncture between these ideals and the actual realities of assessment processes on the ground. Overworked and under-resourced project staff frequently perceive assessment as an additional, and somewhat unnecessary, burden on their time that does not necessarily contribute to their own perceived effectiveness in delivering results. For many organizations, embedding assessment into their operations continues to pose challenges (see Box 6.10).

At the same time, there is considerable methodological innovation in the processes by which decision makers are seeking to measure and understand change. New techniques have evolved, including refinements of existing quantitative methods and experimentation with novel qualitative ones. Organizations are seeking to include a wider range of stakeholders in assessment processes, and there has been

Box 6.10. Response assessment in learning organizations

Assessment is seen as an important learning opportunity by a number of organizations involved in the planning and implementation of policy interventions. It is considered of utmost importance that learning or information generated through evaluations is fed back for better program management. For instance, the World Bank explicitly states, under its strategic objectives of conducting evaluation, the need for "building learning opportunities into evaluations." Similarly, other organizations stress the need to take the feedback of stakeholders on findings, disseminate the results of evaluation through publications, and use findings to improve the performance of the project.

The Joint Inspection Unit (JIU) report on *Managing for Results in the United Nations System* recognizes the importance of learning, but also the limited extent to which this is actually happening. It observes that "in general, there is a lack of coherence in the planning, programming, budgeting and evaluation cycle in many organizations, in particular in the final stage of evaluation, which is not systematically used on the eve of the next budgetary cycle." This suggests that, at present, evaluation findings and recommendations are not being used to their full potential to improve performance. The JIU report suggests that for a results-based management system to be effective, "evaluation findings (must) be effectively used." However, the report does not provide further recommendations on how to improve the effective use of evaluation findings in policy learning.

Organizations such as NORAD, the United Nations Development Programme, the German development agency, the Global Environment Facility, and the Canadian International Development Agency specifically assign responsibilities and detail the processes for follow-up on evaluation findings and their integration into a results-based management system. Guidelines also specify the need to disseminate the findings of the evaluation among interested parties as well as making them available for the public. Within the nongovernmental sector, "critical stories of change"—a method used by ActionAid to document and analyze progress of an intervention—itself serves as a learning document, as the process involves continuous dialogue with the stakeholders at various levels. The document is also available for general public access, to allow wider stakeholder engagement with the findings.

Sources: JIU 2004, ActionAid 2006, IEG 2007.

some progress in developing participatory approaches that empower beneficiaries to determine what matters to their lives and how change should be measured and documented. Apart from beneficiaries, assessment feeds into organizational interaction with a wider range of stakeholders, including clients, implementation partners, donors, and benefactors. Measuring the impact of programs, and documenting change, is a useful way to improve communication with these external stakeholders. It also helps provide greater accountability, allowing organizational performance to be assessed against the expectations of these other actors.

Ultimately, the context within which strategies are being implemented shapes the extent to which assessment of interventions feeds into the decision-making cycle. The implementing agency might be open to a process of reflexive, self-critical learning, in which case the assessment would be seen as part of an evolving cycle of introspection and adaptive management. On the other hand, the assessment might also be conducted by a more defensive and closed organization, which is seeking evidence to justify its intervention but does not actually learn from experience and integrate it into better future planning. There are probably a range of organizational types along the spectrum that these stylized (and polarized) examples represent. What is important is to recognize that the value of any assessment is partly shaped by the attitude of the agency that is conducting the exercise and by its willingness to learn from such a process.

6.5 Conclusion

Section's take-home messages

- In any assessment, it is fundamentally important to identify the nature of the problem that is being addressed, the range of strategies that are available, and the key actors who are able to exert an influence on the issue.
- The value of an assessment process lies is in its ability to progressively improve the quality of intervention strategies.

This chapter has outlined a series of steps that are useful in conducting an assessment of strategic interventions designed to positively influence ecosystems and human well-being. Such an exercise can be conducted to evaluate the performance of previous interventions or to plan future options. In both cases, it is fundamentally important to identify the nature of the problem that is being addressed, the range of strategies that are available, and the key actors who are able to exert an influence on the issue. When looking at past interventions, it is useful to consider not just the options that have been implemented but also other options that might have been feasible but were rejected. It is important not just to assess whether the intervention worked—in the sense of delivering desired outcomes—but also whether it was more (or less) effective than alternatives that might have been pursued.

Similarly, in forward-looking scenarios, a range of possible alternatives need to be considered, and choices need to be made that address decision makers' key strategic objectives. In some situations this is likely to involve accepting trade-offs, in the sense that certain objectives might be achieved through the intervention while others might have to be sacrificed. This is the inevitability of policy choices, and decision makers need to recognize that it is often not possible to achieve progress on all fronts simultaneously. The criteria for choosing among alternatives need to reflect social and political values and realities. Different decision makers, operating in different decision contexts, might have very different rankings among competing objectives.

The value of an assessment process lies in its ability to progressively improve the quality of intervention strategies. In a decision context that values policy learning, an assessment process is likely to yield useful insights into the sorts of strategies that have worked well and those that have been less effective. This can be used to communicate outcomes to stakeholders and to plan future interventions in a cycle of adaptive learning and management. To benefit from this process, decision makers must attempt to create a culture of reflexive learning in order to maximize the influence of the sorts of assessment procedures that have been outlined in this chapter.

References

ActionAid. 2006. Accountability, Learning and Planning System (ALPS). Johannesburg: ActionAid International.

Adams, W., D. Brockington, J. Dyson, and B. Vira. 2003. Managing tragedies: Understanding conflict over common pool resources. *Science* 302:1915–16.

Agrawal, A., and C. C. Gibson. 1999. Enchantment and disenchantment: The role of community in natural resource conservation. *World Development* 27 (4): 629–49.

ASB (Alternatives to Slash-and-Burn Programme). 2003. *Balancing rainforest conservation and poverty reduction*. Policy Brief 5. Nairobi: ASB Programme.

Baland, J.-M., and J.-P. Platteau. 1996. *Halting degradation of natural resources: Is there a role for rural communities?* Oxford: Oxford University Press.

BBOP (Business and Biodiversity Offset Programme Secretariat). 2008. A draft consultation paper for discussion. Available at www.forest-trends.org/biodiversityoffsetprogram/documents/UNEP/BBOP-UNEP-CBD-COP-9-Inf-29.pdf.

Bohensky, Erin, and Timothy Lynam. 2005. Evaluating responses in complex adaptive systems: Insights on water management from the Southern African Millennium Ecosystem Assessment (SAfMA). *Ecology and Society* 10 (1): 11.

Brandon, Katrina. 2002. Perils to parks: The social context of threats. In *Parks in peril: People, politics and protected areas*, ed. Katrina Brandon, Kent Redford, and Steven E. Sanderson. Washington, DC: Island Press.

Chomitz, Kenneth M., Timothy S. Thomas, and Antonio Salazar Brandao. 2004. *Creating markets for habitat conservation when habitats are heterogeneous*. Policy Research Working Paper 3249. Washington, DC: World Bank.

EEA (European Environment Agency). 2005. *Effectiveness of urban wastewater treatment policies in selected countries: an EEA pilot study*. Report No 2/2005. Copenhagen: EEA.

IEG (Independent Evaluation Group). 2007. *Evaluation methodology*. Washington, DC: World Bank.

IUCN (International Union for Conservation of Nature), U.N. Environment Programme (UNEP), and World Wide Fund for Nature (WWF). 1991. *World Conservation Strategy*. Gland, Switzerland: IUCN, UNEP, and WWF.

JIU (Joint Inspection Unit). 2004. *Implementation of results based management in the United Nations organisations: Managing for results in the United Nations system*. New York: United Nations.

Kapoor, Aditi. 2003. Women, water, and work: The success of the Self-Employed Women's Association. In *World Resources 2002–2004*, 198–207. Washington, DC: World Resources Institute.

Leach, M., R. Mearns, and I. Scoones. 1999. Environmental entitlements: Dynamics and institutions in community-based natural resource management. *World Development* 27 (2): 225–47.

Lucas, Nicolás J., Iokiñe Rodriguez, and Hernán Darío Correa. 2008. To change global change: Ecosystem transformation and conflict in the 21st century. In *Policies for sustainable governance of global ecosystem services*, ed. Janet Ranganathan, Mohan Munasinghe, and Frances Irwin. Cheltenham, U.K.: Edward Elgar.

MA (Millennium Ecosystem Assessment). 2005a. *Ecosystems and human well-being: Policy responses*. Washington, DC: Island Press.

MA. 2005b. *Ecosystems and human well-being: Synthesis*. Washington, DC: Island Press.

Olson, Mancur. 1965. *The logic of collective action: Public goods and the theory of groups*. Cambridge, MA: Harvard University Press.

Ostrom, E. 1990. *Governing the commons: The evolution of institutions for collective action*. Cambridge, U.K.: Cambridge University Press.

Panda, Smita Mishra. 2007. *Women's collective action and sustainable water management: Case of SEWA's water campaign in Gujarat, India*. CGIAR Systemwide Program on Collective Action and Property Rights. Washington, DC: Consultative Group on International Agricultural Research.

SAfMA (Southern African Millennium Ecosystem Assessment). 2004. *Nature supporting people: The Southern African Millennium Ecosystem Assessment integrated report*. Pretoria: Council for Scientific and Industrial Research.

Somanathan, E. 1991. Deforestation, property rights and incentives in Central Himalayas. *Economic and Political Weekly* 26 (4): 37–46.

Index